Cattle Ranch

Cattle Ranch

The Story of the Douglas Lake Cattle Company

Nina G. Woolliams

Douglas & McIntyre
Vancouver / Toronto

To all people, past and present,
who have loved these half million acres

Douglas & McIntyre
1615 Venables Street
Vancouver, British Columbia

Canadian Cataloguing in Publication Data

  Woolliams, Nina G., 1947–
      Cattle ranch

      Bibliography: p.
      Includes index.
      ISBN 0-88894-355-5 (pbk)
      1. Douglas Lake Cattle Company - History.
      2. Ranches - British Columbia.   3. Nicola
      Valley, B.C. - History.   I. Title.
      FC3845.N52W65   971.1′41   C79-091118-3
      F1089.N52W65

Maps by Nina Woolliams and Jaclynne Campbell
Photograph opposite title page courtesy Public Archives of
    Canada, National Film Board collection
Jacket design by Jim Rimmer
Typesetting by The Typeworks, Mayne Island
Printed and bound in Canada by
D.W. Friesen & Sons Ltd.

ACKNOWLEDGEMENTS

My first recognition goes to the many who have lived and laboured for Douglas Lake, then and now, for they kept the ranch's legends and annals so alive that the story took hold of me within the first year of moving to this fabulous ranch.

Next came a group of people who brought me through the transitionary period between researching for the fun of it, and researching for a book. They include the late Talbot Bond, the late Joseph Guichon, Mrs. Nellie Guitteriez, the staff of the Kamloops Court House, especially Mrs. Joy Smith, and Mrs. Mary Balf, curator of the Kamloops Museum Association.

Brian K. de P. Chance gave excellent assistance from this time forward in the form of masses of old records, photographs and memories. From the first draft on he was eager to proofread, correct and explain once more and has thus been to me an invaluable and patient authority and friend.

C.N. Woodward, present owner of Douglas Lake Cattle Company Ltd., gave the infant project his corporate blessing.

My research then branched out in every direction and I received assistance from 22 archival repositories at home and abroad (in particular the Provincial Archives of B.C. at Victoria), four local newspaper offices, two British county councils, 11 provincial ministries for British Columbia, two federal ministries, four corporations, and well over a hundred individuals living in three different countries. Of those, I wish to extend personal thanks to Mrs. Elizabeth Abbott, Miss Margaret Abbott, Mrs. J. Allan, Mrs. Elsie Beairsto, Humphrey Beak, Mrs. Ernie Blades, H.M. Boyce, Mrs. Phyllis Brewer, Alex Bulman, J.W. Cartwright, Mr. and Mrs. Richard Chamberlain, Rex Chapman, Robert Clemitson, Hon. J.V. Clyne, William G. Coventry, John Roy Edward Douglass, George Farmer, Mrs. Betty Farrow, Mike Ferguson, Chay Gilchrist, Joe D. Gilchrist, Mr. and Mrs. Alex Gillespie, Mrs. Lawrence Graham, Raymond Graham, Gerard Guichon, Mr. and Mrs. Nigel Gyles, Mrs. Luella Hare, Harry

J. Hargrave, Harley Hatfield, Reid Johnston, Eddie King, Joseph W. Lauder, Dr. Douglas Leechman, Leslie Leighton, Francis V. Lumb, Micky Lunn, Miss Jean McDonald, Miss Jessie McDonald, Mrs. Lina McKay, Angus McLeod, Mr. and Mrs. Robert Marston, Miss Betty Munro, Mr. and Mrs. John O'Reilly, Tommy O'Rourke, Mrs. Harriett Paul, Horace Plimley, Mrs. Frank MacKenzie Ross, Mr. and Mrs. Ed Ruman, Walter Russell, Joe Sledge, Douglas Carson Smith, A.F. Sproule, Miss Sophie Steffens, John M. Tennant, Donald C. Tuck, Russ Turnbull, Des Vicars, Norm Wade, Donald Waite, Curtis Ward, Dr. Donald Watkins, Mrs. John J. West, and Lee Williams. A reckoning of outgoing correspondence totals over 700 letters.

The selected bibliography following the text credits the most extensively used sources around which I built the story. The captions below each photograph credit their source.

Professor Gordon Elliott of Simon Fraser University gave help of an editorial nature.

No nonfiction writer is an island, and without the substantial assistance of all these people, *Cattle Ranch* would still be a dream. I thank each one.

Throughout every stage, however, from conception to delivery, the greatest help has come from my husband Neil. I thank him for taking the successive roles of mentor, critic, shoulder-to-cry-on, ranching encyclopedia, cajoler, advocate, proofreader and expectant father during the eight-year gestation period of *Cattle Ranch*.

Nina G. Woolliams,
Douglas Lake, 1979

CONTENTS

MAPS

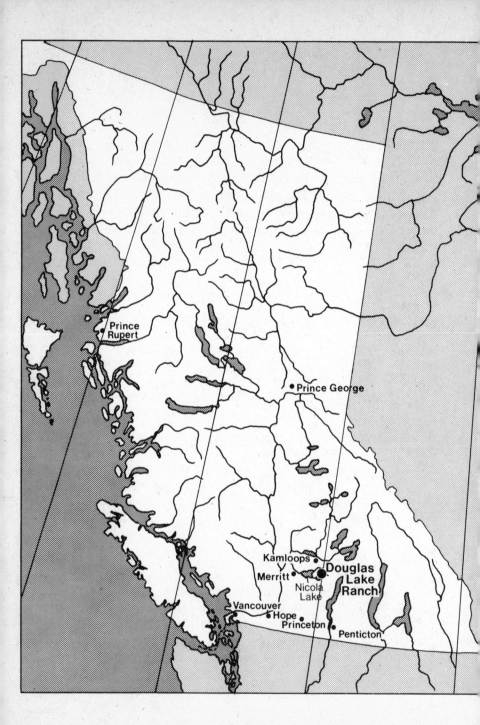

Prince
Rupert

Prince George

Kamloops
Merritt
Nicola
Lake
Douglas
Lake
Ranch
Vancouver
Hope
Princeton
Penticton

INTRODUCTION

The Air Canada jet lifts out of Toronto and heads for Vancouver. It edges along the Great Lakes and flies over evergreen forests shimmering with water. The trees thin out and the prairies spread below as postage stamps of yellow-hued grains. Suddenly a wall of jagged mountains soars high, grey and forbidding. The pilot beams in on the aeronautical landmark of Enderby and towards the air path known as Vector 304. Streaked with rivers and splashed with lakes, the mountains continue in lesser and lesser series to the west coast and Vancouver.

Now for a short time the never-ending carpet of conifers ends. Below is British Columbia's Interior ranch land, and here, captured for a moment, is Canada's largest cattle ranch. Douglas Lake Cattle Company holds such a title because of the extent of its land—163,000 deeded acres controlling more than 350,000 acres of Crown grazing—and the number of cattle it carries—a wintering herd of 11,000 head of commercial Herefords.

The air is clear. Startlingly green haylands and flashing lakes leap up from the backdrop of faded brown velvet grasslands. Around its edges, dots of timber rise into a semicircle of mountains to the east: mountains stippled green, blue and violet.

A closer look brings these features into greater relief. Through the heart of British Columbia's southwest Interior flows the Fraser River system; into the Fraser flows the Thompson River; into the Thompson flows the Nicola River. The upper stretch of the Nicola gathers run-off from a half-million-acre plateau of bunchgrass and timber. A semicircular mountain range to the east separates the Nicola Valley from the Okanagan Valley. The southwesterly mountain range which includes the Coquahalla Pass blocks off Hope on the Fraser River. Several cattle outfits bordering Nicola Lake trace a boundary to the west and north.

The half-million acres of this entire upper Nicola River watershed area belong to Douglas Lake Cattle Company, making it not just Canada's largest ranch but one of the world's greatest, for the watershed possesses all the geographical features necessary for cattle ranching: spring, summer and fall range; flat land for cultivation; abundant water. The Home Ranch on the shores of Douglas Lake is its centre, to which the cattle naturally drift, for here at 2,600 feet is the area's lowest elevation.

In ranching a balance must be struck between the size of the cattle herd and the feed available: spring, summer and fall pasture and the hay tonnage grown for winter. At Douglas Lake such a balance presently exists with a wintering herd of 11,000 head. For the herd to increase, the productivity of the entire ranch must increase.

This Nicola Valley comprises the tip of a desert that stretches north from Mexico. Because most of the annual 10 inches of precipitation falls in spring, at the height of the growing season, the land resembles open parkland rather than desert. Even so, desert plants such as prickly pear cacti flourish on the more barren, lower elevations of southern exposure. The drought-resistant bunchgrass growing here is in prime condition, yet between each 2- to 3-foot-high clump is bare earth: space for each delicate root structure to gather moisture. Thus for every 35 acres of this drybelt, from the highly productive spring and fall ranges to the jackpine jungle and swamp meadows supplying summer grazing, there is just one head of beef.

These cattle must be hardy, for they cover much ground in the course of a day's grazing. They spook easily; a man on foot causes them to scatter with a kick of their heels. They suffer many predators: bears, cougars, coyotes, rustlers. Even a raven or a bald eagle can pick the eyes out of a newborn calf. One to two per cent losses of mature stock annually are normal. Only the strong and lucky survive the climate and the predators, and they adapt to these conditions and prosper.

For every 300 head of cattle at Douglas Lake, there is one person on the steady crew, either a farmer growing winter feed, a cowboy tending and moving stock, or a cook. These men follow a work pattern that covers 12 months.

During the three months of winter when the mercury often drops to minus 30 degrees Fahrenheit and the sun appears for just eight hours each day, a foot or more of snow seals up the grasslands. Douglas Lake's remuda of 200 saddle horses paws through this snow for feed, but cattle, not having the pawing instinct, must be fed hay when the snow gets deep or crusted. This winter feeding generally starts during the week of Christmas. From then until late March, the herd consumes around 100 tons of hay each day; by the end of winter it will have eaten up to 11,000 tons.

In February, the bred heifers that are about to have their first calves move to the holding fields around the Maternity Barn at English Bridge and calve out under 24-hour surveillance. The next month the cowboy crew begins calving out the cows on range—riding daily through the herd to help with difficult births. Five thousand calves is an average annual crop.

Turnout from the winter feedgrounds to the spring ranges of open bunchgrass commences in mid-March and continues through April (with the cattle being moved higher as spring reaches the upper elevations). Even though the cowboys have sprayed all the cattle against pests, riders constantly patrol particular fields to help cattle stricken with ticks: parasites that can cause paralysis and subsequent death if they are not picked off quickly. The cowboy crews move out with the cattle in late April, taking their strings of saddle horses, their bedrolls and little else with them. They have a month of branding ahead.

Taking advantage of cool, early mornings when the sun rises at four o'clock, a full crew of 25 cowboys drives the cattle to the high summer grazing in the timbered mountains fanning out from the open bunchgrass lowlands. The bulls—300 of them—go with the cows now; 20 cows to each bull. Once their charges are in the high country, the cowboys settle down to long, hot summer days of keeping stock scattered and salted, shipping dry cows that lost their calves early on, breaking young horses, and repairing fences.

From the moment the farm crews feed out the last lick of hay in spring, they start preparing the land for growing a year's forage crops of 11,000 tons of hay plus silage and grain. Spring plowing, new seeding, fertilizing by air, cleaning irrigation ditches: all these tasks precede irrigation of the haylands either by flooding or by sprinkling. Two crops mature at the Home Ranch and Chapperon, while higher up at Norfolk and Minnie Lake, one crop requires the whole season to reach the same ripeness. Four-ton hay loaves dot the hayfields as they dry; then the farmers mass them near the winter feedgrounds. The ranch hands' air-conditioned tractors make the cold and windy work of spring, the mosquito-infested haying season, and the months of winter feeding more bearable than in the past, even though the hours are still long.

As the first, newly opened gates at the higher elevations allow the cattle to start drifting down from the snow line, the cowboys drive the long yearling steers (18-month-old males) in from closer range for one of the most important sales of the year—the Panorama Sale. Held in conjunction with other Interior ranches, this sale sets the budget for the following year's operation. Stock that is sold later in the fall includes grass-fat heifers (females) for slaughter or feedlot, veal or other calves, culled cows and bulls. More than 4 million pounds of beef are sold during the 12 months, enough to feed a town of more than 20,000 people for a year.

Most of the herd drifts slowly homeward, and the cowboys sort the cows and calves according to the sex of each calf. Two big drives from the Minnie Lake area to the Home Ranch—2,500 cows and their 2,500 heifer calves, and 2,500 cows and their 2,500 steer calves—occupy a week in November. Ten miles is a normal day's drive.

Progeny testing takes place: the cows that have brought home poor calves must leave the herd. The cowboys wean the calves from their mothers and put them on separate feedgrounds farther up the valley. Pregnancy testing of the cows eliminates the late calvers and open stock (barren cows), thus whittling the herd down before winter feeding. The calves are fed hay now, while the older stock stay out as late as possible to rustle what grass the snow leaves uncovered.

There are additional members on the total Douglas Lake crew who are not involved directly in the cattle operation: they include the dairyman, who supplies fresh milk for orphan calves and for employees; the gardener, who grows sufficient vegetables for the year; the man who tends the pigs; and the general storekeeper who sells groceries, gloves, rope, etc., mans the gas pumps and sorts the mail. Then there are the office staff, the two mechanics, and the fencing foreman, who with an Indian crew builds most of the new fences, gates, yards and corrals on the ranch and who keeps in repair the 700 miles of older fences.

Another crew cares for and trains the 100 head of registered Quarter Horses. The Quarter Horse barn at the Home Ranch has been home to two world champion cutting horses, Peppy San and Stardust Desire, and to many Canadian champions.

A business executive, C.N.W. (Chunky) Woodward, chairman of Woodward Stores Limited, solely owns this block of real estate which is the largest privately owned ranch in Canada. About twice a month, he flies from Vancouver to land at the ranch airstrip. He takes part in ranch work, and goes riding, fishing, or hunting in season. The ranch's manager, E. Neil Woolliams, is a university graduate full of innovative ideas. He has spent 19 years at Douglas Lake, 12 of them as manager. Woodward and Woolliams meet regularly to discuss ranch affairs; a love for this land guides both men in carving out its future with care.

There have been only three other managers and three other sets of owners in the ranch's almost century-long history. But the history goes back further to the early pioneers and the native people who came here. They are the ones who first appreciated the beauty and the value of this rolling rangeland, its ability to raise cattle, and its potential to convert good hard grass into salable pounds of beef.

Ranching is an unpredictable venture. Nothing about it is constant—neither the animals, the weather, the people, the machinery, the land, nor the amount of government interference. This unpredictability makes for a business that is tough and challenging both physically and economically and that attracts and holds only those who have spirits to match. It has always been so, even though its make-up has changed often.

These many changes have written Douglas Lake's story in every grove of trees, every bench of rangeland, every red-roofed barn and every field of hay within its half-million acres. May that story go on without end.

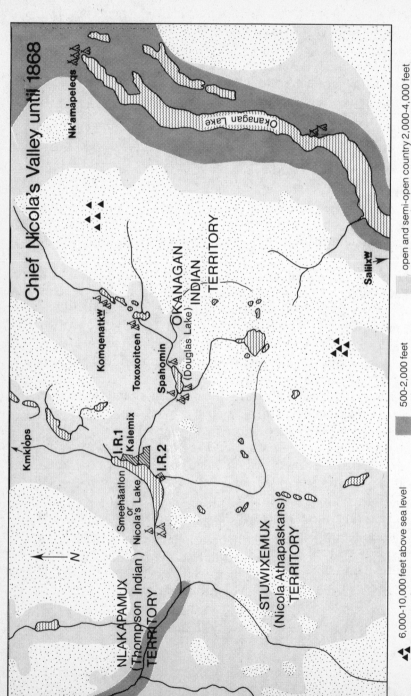

Chief Nicola's Valley until 1868

Nk'amápeleqs

Okanagan Lake

Komqenatkʷ

Toxoxoitcen

Spahomin

(Douglas Lake)

OKANAGAN INDIAN TERRITORY

Saliixʷ

Kmkʼóps

Smeehaatlon or Nicola's Lake

I.R.1 Kalemix

I.R.2

NLAKAPAMUX (Thompson Indian) TERRITORY

STUWIXEMUX (Nicola-Athapaskans) TERRITORY

N

6,000-10,000 feet above sea level

forested land 4,000-6,000 feet

500-2,000 feet

Indian Settlements

open and semi-open country 2,000-4,000 feet

Okanagan Indian Reserves of 1868 (I.R.)

PROLOGUE

During the reign of Chief Rolls-over-the-Earth (*Pelk'am-ólaxʷ*), the Okanagan people tired of warring ceaselessly and the chief led 400 of his people north to an uninhabited plateau land. The Okanagans settled into the peaceful, rolling landscape, erecting their summer lodges on the shores of Head Water Lake (*Komqenatkʷ*). The river springing from Head Water Lake was one of many supplying Shuswap Lake.

South of Head Water Lake, over the height of open land that sent water southwest instead of north, lay another lake (*Toxoxoitcen*). Its spillage flowed into a lake soon named *Spahomin*, for around its shores were willows, the young branches of which the Okanagans shaved and smoothed to use as the hoop of their dipnet. The name *Spahomin* came from the shaving done at this place.

From *Spahomin*, a river flowed northwest to a new-moon-shaped lake (*Smeehāatlon*). Where river and lake met stood Willow Bushes (*Kalemix*), a settlement of 60 people. An offshoot of an Athapaskan tribe (*Stuwixemux*), these people had left the Chilcotin as a warring party that lost its way. At the mouth of the new-moon-shaped lake more Athapaskans camped.

Farther west along larger rivers lived a third tribe, the Thompson Indians (*Nlakapamux*), some of whom were already marrying into the Athapaskans and sharing their hunting grounds.

The Okanagans summered at Head Water Lake but initially wintered at the Head of Okanagan Lake (*Nk'amápeleqs*) because the bitter cold and drifted snows were less intense there than at their newfound home.

Chief Rolls-over-the-Earth occasionally went buffalo hunting on the eastern plains and during the first decade of the nineteenth century, he met there white men, Finan McDonald and Legacé, North West Company voyageurs.

Chief Rolls-over-the-Earth, thrilled with this exciting encounter, described his meeting with the strangers to his people and to other tribes. He became a celebrity and travelled far as a storyteller. However a Lillooet (*Stlat'lémux*) chief doubted Rolls-over-the-Earth's word, publicly called him a liar, and fatally wounded him with an arrow.

Chief Rolls-over-the-Earth died at the end of an era: when the Indians could migrate and retreat and advance with only the willingness or unwillingness of a neighbouring tribe to influence their movements.

The Europeans who were exploring the Northwest at this time thought that the Indian people were settled in territories and did not realize that they would have continued as nomads if it had not been for the establishment of fur-trading posts around which their lives soon centred.

Walking Grizzly Bear (*Hwistesmetxéqen*), the son of Rolls-over-the-Earth, became chief of the Okanagans during this period when the white man was thrusting his nineteenth century civilization upon the Indian.

In the spring of 1813, before 25-year-old Chief Walking Grizzly Bear had moved from his winter quarters at the Head of Okanagan Lake to summer quarters at Head Water Lake, Montigny of the Pacific Fur Company entrusted him with the safety of some trade goods. After transporting the winter's furs to Astoria, the fur trader returned and rewarded the chief's reliability with ten guns, ammunition, pipes, tobacco and vermilion. He also renamed him Nicolas: Nkala on the tongues of the Indians and Nicola to the next generation of Europeans.

Chief Nicola, the most influential chief of the Interior of New Caledonia, educated his braves in the use of the firearms. Gathering allies from neighbouring tribes, he made war on the Lillooet Indians who had killed his father. Nicola celebrated his revenge with a feast of elk for his allies, piling high the resultant elk antlers and bones near the crescent-shaped lake that became Nicola's Lake. The valley bottom and the plateau land beyond in all directions became Nicola's Valley.

Fur traders from Thompson's River Post at *Kmklóps* encouraged Nicola and his people to abandon all activities in favour of trapping, and the Okanagans' language, unwritten as it was, soon embraced French and Chinook words. But gradually the Indians' independence, dignity and way of life deteriorated as they further depended on trade goods and accepted the white man's way as theirs.

The extinction of beaver and marten through heavy trapping coincided with the decline of the European fur market and with the arrival of the gold miners.

This influx of miners in 1858 occasioned the founding of the Crown Colony of mainland British Columbia. (The British Crown had granted the colony of Vancouver's Island to the Hudson's Bay Company in 1849.) The miners' sudden arrival so alarmed the Thompson Indians that war was imminent. Only the word of Nicola, the powerful chief who eventually took 17 wives, his supreme influence, and his respect for Queen Victoria, succeeded in quelling his neighbours.

After 74-year-old Chief Nicola died, his nephew, *Chillihitzia*, became chief. Starvation, confusion and strife filled the early years of his rule, the era of the gold miners of the Fraser River and later of the Cariboo. Chief *Chillihitzia* was not a healthy man, and more than once his people prematurely prepared for his funeral.

En route to the goldfields, miners occasionally left the more travelled Okanagan and Fraser River routes and journeyed through Nicola's Valley. Within a decade, white men began settling around Nicola's Lake.

Peter O'Reilly, the Indian reserve commissioner for British Columbia, noticed their arrival and in August 1868, on the instructions of Joseph W. Trutch, chief commissioner of lands and works, reserved what he thought was 1,080 acres at Willow Bushes for *Chillihitzia*'s tribe of 150 people and their 20 head of cattle and 130 horses. He also set aside a fishing reserve of 80 acres farther south on Nicola's Lake.

What Commissioner O'Reilly estimated as 1,080 acres was in fact three quarters of this figure, roughly 10 acres per family. Ten acres, when *Chillihitzia* and his people had been accustomed to vast areas! It was going to take many years for the Indian people to understand that they were to stay confined within their reserves; for the whites to realize that these reserves were too small to sustain the Indians; and for the Indians to accept coexistence with the whites.

Under three chiefs the Okanagans' way of life had been transformed. As their tribal territory became occupied by opportunist gold miners who were realizing that a more certain wealth lay in homesteading, *Chillihitzia*'s people gradually turned their hands to new endeavours.

CHAPTER ONE

Old Douglas was evidently the oldest bachelor among the residents of that area. In the years prior to his marriage he was always giving advice to the younger bachelors as to what kind of a girl they should pick for a rancher's life—they should choose from among those in the country rather than the city. His experience evidently proved his argument.

Lawrence P. Guichon to Brian K. de P. Chance,
29 January 1956

The Okanagan Indians had spaced their dwellings along the banks of the gently flowing Upper Nicola River, between the groves of aspen and cottonwood. Most of their summer lodges and keekwillies (below-ground winter lodges) were in a natural meadow at the western base of Spahomin Lake, out of which the river flowed.

Around the 3½-mile-long elliptic lake, above the fringe of willows, grassy slopes swelled away, devoid of trees where they faced south and dotted with virgin stands of Douglas fir where they faced north. From a distance the dry brown soil seemed clothed in ochre-coloured velvet—the tall native bunchgrass, cured by the summer's sun.

Few white men had penetrated here, 14 crow miles east of the bustling young settlement of Nicola on the western end of Nicola Lake. Clement Cornwall of Ashcroft Manor had been the first in 1863 when he led a grouse-hunting expedition through.

Beyond the willow barrier to the east end of Spahomin stretched a natural amphitheatre of bunchgrass, bounded by trees and the river to the west and north, trees and the lake to the south, and more rolling hills beyond. A bowl-shaped lake nestled into the eastern edge of the enormous meadow.

It was this amphitheatre and the shore of this small lake, which he named Round Lake, that John Douglas, Sr. decided to homestead in the fall of 1872. He staked out the

320 acres that the British Columbia Land Ordinance of June 1870 allowed him to claim as an agricultural land preemption: acreage occupied before survey and purchase. He rode 100 miles to the magistrate's office in Lytton to record his half-section preemption on 17 September 1872.

His nephew, John (Jack) Douglas, Jr., recorded his own 320-acre preemption two weeks later, on 2 October. His stakes ran through part of the natural meadow at the west end of Spahomin Lake. The Indians were most uneasy to see him settle in their midst.

John Douglas, Sr. and John Douglas, Jr. were Scottish immigrant pioneers. Old Douglas, so named to differentiate him from his nephew, was just 43 that year, though there were already grey whiskers in his thick black beard. He suffered from a tubercular chest that he hoped the dry climate would help; this respiratory condition had dissuaded him from staying with his dairying brother William on the marshy land north of the state of Oregon.

Douglas decided to raise beef cattle, though earlier settlers to the more travelled part of the Nicola Valley, such as the Mexicans Pancho Guitteriez, Raphael Carranzo, Blass Leon and Jesus Garcia and the South American Anton Godey, had not made the same choice. As packers of freight from Yale to the Cariboo goldfields they had found the extensive grasslands around Nicola Lake in 1865 when looking for winter pastures for their trains of horses and mules.

The packers were not the first to preempt land in the valley, though. That honour fell three years later to two English sheepmen, Edwin Dalley and John Clapperton. In that year of 1868 the Nicola Valley became peopled from one end to the other. William Charters, the first of four brothers to come out from Liverpool, opened a trading post at Forksdale at the confluence of the Nicola and Coldwater rivers. Florein Mickle, a blacksmith, and his teamster brother, Wheeler, settled at Quilchena, where the creek of that name flows into Nicola Lake. Two drovers known as Fish and Snyder brought their pack oxen to winter next to the first Indian reserve that Commissioner O'Reilly had established, where the Upper Nicola River flows into Nicola Lake. And the Moore brothers, Samuel and John, the first to bring cattle into the valley, settled at the north end of Nicola Lake, the crescent-shaped body of water that dominates the geography and life of its valley.

The settlers had come in diverse ways. The Mickles had been part of the famous Overlanders' party of 1862. The Moores had also travelled overland, via forts south of the border, in 1860. Others had travelled over the plains of America in covered wagons, on the route of the California gold rush, via Cape Horn in sailing ships, via the Isthmus of Panama and San Francisco, via the Fraser and Cariboo gold rushes. They had come from Ireland, Scotland, England, Germany and Ontario. They were brothers, parents, sweethearts, friends and strangers to the original settlers. They came on horseback and on foot: the wagon maker and the miner, the businessman and the grain farmer, the shepherd and the rancher.

There were many ways John Douglas could acquire cattle. Although the numerous descendants of the longhorns that the Spaniards had brought to Mexico in the sixteenth century had spread north, they had not reached British Columbia. Earlier in the nineteenth century, the Hudson's Bay Company had brought the rough-coated brown, red or white Durhams to the northwest coast as a meat source for their employees. These had multiplied sufficiently to use as trade with the Indians and white settlers. From the eastern American states to Oregon and California came more Durhams from established herds. Thus, like the Moore brothers, John Douglas could drive in a herd from Oregon. Like John Gilmore at Nicola, he could buy from the Indians. Or he could buy from better-established white neighbours.

It would be two and three years before Douglas would receive income from his herd. Where would the markets be by then? The Cariboo market was in decline and many gold diggers had abandoned their claims to become farmers and fishermen, ranchers and restauranteurs.

The coast settlements of New Westminster and Victoria had replaced the goldfields as British Columbia's largest market, but to drive cattle from the Nicola Valley to the coast was a

14

horrendous undertaking. The success or failure of Nicola ranchers was based upon the stock routes out of the valley.

Of the three trails leaving Nicola, the first went 50 miles north to Kamloops, far from coast markets.

The second, which went west to Cook's Ferry on the Fraser River, had been blazed earlier that summer in preparation for widening and straightening. Mostly it stayed on the north bank of the Nicola River, but as that river neared its confluence with the Thompson River the valley narrowed and the trail crossed back and forth. At Cook's Ferry, the bridge Thomas Spence had built seven years earlier took travellers to the southeastern bank of the Thompson. This route took its users miles north of their destination.

The third trail, which led to Boston Bar on the Fraser River, struck through high mountains circling the Nicola Valley to the south. Although it was the least devious and the Indians claimed it to be "the best travelling," its rocky miles offered scant feed for any cattle drive. Both these western routes then followed the rocky shore of the Fraser River to Yale, from where steamers would carry the cattle down to market.

Victoria's newspaper, the *Colonist*, had first pinpointed this difficulty the previous year. "The citizens of Victoria and New Westminster and the people of Burrard Inlet pay $100,000 a year for beef. Every dollar of this goes to our friends in America." The *Colonist* suggested that the government build a sleigh road from Lillooet via Pemberton Meadows to Burrard Inlet at the cost of $5,000, although this route would help only the ranchers living north of Lillooet.

But marketing was not yet a problem for John Douglas who was still acquiring stock: yearling cattle for $30; two-year-olds, $45; three-year-olds, $50; hay, $10 per ton. For his own food he was paying per pound 12½ cents for pork, 50 cents for butter, 2 cents for wheat and 1¼ cents for potatoes.

Douglas's ranch was to carry a thousand head of cattle. In every direction, the natural barriers of his land prevented his cattle from wandering. In these days of unfenced range, this was important. He built a simple log cabin in a clearing in the willow brush between the two lakes, and from this base learned the annual pattern of nature in his part of the Nicola

Valley: rushing streams and mud after the first thaw; bunch-grass shoots and flocks of waterfowl appearing in April; spring winds; June rains; mosquitoes accompanying the 80-degree heat of long summer days; sunburned August ranges; September flies and the first killing frosts; Indian summer days and gold and orange leaves on the willows in October; bare Novembers with the first high snowfalls; winter's permanent arrival in December with frozen lakes, and skiff after skiff of fluffy snow; pindrop calm, short winter days of freezing intensity and crunchy snow sealing up the ground until March.

Settlement at Douglas Lake was slow. Augustus Bercie homesteaded upriver from Old Douglas in July 1873. His land centred on a wild meadow through which wandered the Upper Nicola, and because of his untidy habits the meadow became known as the Boar's Nest. Many years went by peacefully for Old Douglas, Jack and Bercie, years of building corrals, cutting wild hay, and herding cattle.

In 1873, the Moore brothers drove their cattle to Victoria over a new route that first went south to Princeton and then down the Dewdney Trail to Hope. The cattle arrived in Victoria in poor shape. Their feet were sore from the rocky road and they had lost a lot of weight, for feed had been scarce. This route was obviously not the answer. But the drive revitalized discussion on the need for a good wagon road that ranchers could use as a stock trail to the coast.

It was the *Colonist*'s view that the province should first build less expensive stock trails which could later be made into wagon roads, rather than spending up to $4,000 right away for a good wagon road. The newspaper felt that with British Columbia already pushing Ottawa for a transcontinental railway, pressure for more funds might be too much.

By March 1874 the Interior settlers were getting desperate, and 91 of them sent a petition to the lieutenant-governor:

The undersigned settlers of the Kamloops, Okanagan, Nicola and Cache Creek Valleys, beg to petition Your Honour, for the construction of a road from the south end of the Nicola Forks, up the Coldwater Valley to the summit of the Coquahalla, thence down the Coquahalla to Fort Hope. The distressed condition of the stock-raisers of the district, owing to their having no outlet by which they can drive to the now almost only beef market in the Province, together with the fact that the cattle ranges are becoming overstocked and destroyed, we trust will induce you to make some efforts for our relief.

Robert Beaven, chief commissioner of lands and works, instigated explorations of three possible routes over the mountains from the Nicola Valley to Hope. The one via the Coquahalla Pass took in "a very bad, rocky slide formed by immense fragments of granite." One via Princeton offered a mile of precipices and the last via Otter Valley afforded the most feed and the least treachery, but was the longest.

Being the shortest, the Coquahalla was the most attractive route, although one party had been forced to abandon their study in late April "as the snow was getting pretty deep. . . . July would be soon enough. . . . The snow will not be off sooner." It was also found that the Coquahalla was not just 68 miles long but closer to 80, and that in addition to clearing, blasting, cribbing and walling, the contractors would have to build 750 feet of corduroy, 106 culverts, and no less than 60 bridges. For much of the distance there was no cattle feed and because so much snow would fall on it, the trail could never become a year-round wagon road. Even so, work went ahead on the 6-foot-wide trail.

Old Douglas was very interested in the surveys being carried out in the valley in 1874 when John Jane and E. Stephens, two Royal Engineer sappers, surveyed the 41 preemptions by settlers along the route of the proposed wagon road from Kamloops to Cook's Ferry via Nicola.

Douglas's interest sprang from the fact that once preemptors received a certificate of improvement, they could apply for Crown Grants, and if granted, could begin the process again by preempting more land. If he had been able to have his first land surveyed and thus had received the Crown Grant, he could have saved the $200 that he had spent to secure the land lying east of his first parcel. Instead, he was buying the second in instalments. Old Douglas had his eye on other parcels too, but the government allowed a person only one preemption at a time.

Living so far off the beaten track, Old Douglas, his nephew

Jack, Bercie, and three French brothers named Guichon who lived at Mamit Lake had to await the next surveying trip through the valley, which was not until 1877 when Edgar Dewdney surveyed Douglas's amphitheatre. For the 277 acres Douglas paid the usual price for preempted Crown land—a dollar an acre.

Also in 1874, government contractors began work on the most westerly 40 miles of the Kamloops-Cook's Ferry wagon road. A school district was formed for the children of those settlers living at Forksdale and nine miles west. An Ontario sawyer, George Fensom, harnessed the power of the Nicola River and built the valley's first sawmill. Between the crude log and hand-hewn timber cabins rose smart houses built with mill-sawn timbers.

In the rapidly growing village of Nicola in 1876, the Reverend George Murray, a Nova Scotian, built the Presbyterian Murray Church, which became the parent of all churches in the Interior and a base for many other denominations.

At Douglas Lake, the Reverend Mr. Murray's brother, Hugh, settled on Murray Creek, a tributary of the Upper Nicola. Ronald and George McRae came from Scotland and settled between Old Douglas and Bercie. A Frenchman, Napoleon Sabin, preempted a flat meadow bordering the south bank of the Upper Nicola.

New settlers arrived thick and fast in the mid- and late '70s to dot the vast Nicola River watershed with cabins and cattle. A Yorkshireman, Byron Earnshaw, his Indian wife, Shinshinelks, and their children, Harry, Lavina and Minnie, settled 14 miles southwest of Douglas Lake. Their nearest lake in that high plateau country became Minnie Lake. Patrick Kilroy, a Lytton butcher, became absentee owner of more Minnie Lake land. Joseph Dixon Lauder and his wife, a sister of the Moores, settled up Lauder Creek not far from the Douglas Lake trail to Nicola. Another Moore sister and her husband, Robert Scott, settled at Rockford, halfway between Nicola and Stump Lake. Thomas "Pike" Richardson, an American from Pike's Peak, Colorado, located at Chapperon Lake and eventually owned 5,000 acres there.

Wheeler Mickle ran a twice-monthly stage between Cook's

Ferry and Nicola, transporting freight, mail and passengers, and soon stopping houses opened up along the route. Previously, groceries and dry goods had come by pack train once a year from Victoria, when a gala would be held in Nicola.

Work was in progress on the road to Kamloops; the *Guide to British Columbia* tersely advised, "Can bring a waggon with light load across from Kamloops to Nicola Lake, if you take a guide, an axe and a spade."

By now, Douglas owned around 700 acres and ran almost that many cattle, which had taken him less than five years to accumulate. In spring, summer and fall they ranged on the Crown land south of Round Lake. In winter he would move them back to his home place, cutting out any of his neighbour's cattle that might be there, and if his stock could not rustle enough grass through the three snowy winter months, he could feed them the little wild hay he had cut during the summer.

For five years, Jack Douglas had been living among the Indians at the south end of Douglas Lake without realizing how much the Okanagans considered his presence an intrusion on their privacy and their land. One early winter's day in 1877, a Spahomin Indian named Chinook Tom stepped off his horse in front of Jack Douglas's log home. A heavy axe rested on his shoulder. When Douglas came to the door, Chinook Tom said he was on his way to cut firewood, but that if Douglas felt like moving off his land, that would be fine.

Jack hurriedly gathered his belongings and loaded them onto a pack horse, saddled another, mounted it and spurred it east, up the north bank of Douglas Lake. He returned later with help to round up his 300 head of cattle and moved them next to his uncle's place.

Luckily for Jack, that winter of 1877-78 was very mild, for he did not have time to build himself another cabin before winter set in, and instead dug a large hole in the ground and covered it with a pole frame and cowhides, Indian style.

With the coming of spring, Jack built himself another log home at the east end of the lake, north of his uncle's. But this was no permanent solution, for he held a preemption for his land at Spahomin and had made many improvements preparatory to receiving his Crown Grant. An even greater problem was also involved.

A battle had been raging on behalf of the Indians since 1871, when British Columbia had confederated with Canada. The federal government was appalled that British Columbia thought 10 acres per family sufficient for their Indian reserves compared with 80 acres in the rest of Canada, especially when British Columbia allowed a white man to preempt twice as much east of the Cascades and four times as much west.

The Indians, who had not remained on the reserves but lived at their habitual camping grounds, were becoming hostile and awaited only the word of Chillihitzia, the chief of the Okanagans, to start a full-scale war against the white settlers. But Chillihitzia had great faith in Queen Victoria and her laws and sought a peaceful solution.

In September 1878, Gilbert Malcolm Sproat came to Douglas Lake as Indian reserve commissioner to negotiate with Chief Chillihitzia for the establishment of further reserves.

The first issue was that of Jack Douglas's land at Spahomin. Neither Douglas nor the Indians were willing to abandon their claims to the land, so Sproat brought in a compromise. Jack Douglas received "free of cost from the Provincial Government, a Crown Grant for the unoccupied Crown land bounded on the east by Hugh Murray's preemption and on the south and west by the Upper Nicola River" just north of his uncle's place.

At Douglas Lake, Chillihitzia had built a "small keekwilly house and small corral... on a portion of the land assigned to Mr. Douglas Jr." Also there was "the solitary grave of one of Chilliheetza's sons who died a few years ago [and] another grave just off the waggon road as you leave the head of Douglas Lake." On the strength of these graves and house, Chillihitzia asked for all the shoreland around Douglas Lake.

The commissioner offered the chief compensation for the loss of his keekwilly and gave him time to move the graves, but absolutely refused to reserve the entire shoreline of the lake for the Indians. This "so affected Chilliheetza that he shut himself up in his house, but the mass of the Indians were satisfied with my decision... and Chilliheetza afterwards acquiesced in a friendly way."

Chief Chillihitzia reached an agreement with Sproat on two

separate sites around Douglas Lake. These reserves totalled almost 23,000 acres, but Sproat was not yet finished, and now travelled to Chapperon Lake.

There, he stopped to talk with two of the white settlers, Pike Richardson and newcomer Laurent Guichon. Laurent, the eldest of the brothers who had been at Mamit Lake, had left his native France in 1857. He had reached California towards the end of its gold rush, but had been among the first to the Cariboo in 1860. For nine years he, his brother Pierre, and Vincent Girod had been partners working out of Lytton in a "very extensive and remunerative business in mining supplies and general merchandise from Yale to Quesnelle and Cariboo." They were leading pack train owners and their brother, Joseph, worked for Cataline, another well-known pack train owner.

Sproat wished to create a fishing reserve for the Okanagans on the west shore of Chapperon Lake, and learned from Richardson and Guichon that "this piece would be of much use to the Indians and not in the way of the whites." Sproat added that:

It is no uncommon thing to see 1,000 Indians with their horses there in the spring, from Osoyoos, Okanagan, Similkameen, Kamloops and Nicola. They have a race course, and for so many horses, and for their encampment and fishing purposes and feed for animals, require a considerable space. The reason for the above periodical gathering of the Indians is that in Chapperon Lake a trout of large size and firm quality takes bait earlier in the spring than in any other lake in the southern interior of the Province, and is not, therefore, a gathering merely for sport, but for obtaining a welcome supply of much prized food after the privations of the winter. This reserve . . . is in a district where nothing can be cultivated owing to its elevation.

This reserve, later known as Ginny's Flat, along with two others nearby, the two at Douglas Lake and the two created in 1868, gave the 160 Okanagans in the Upper Nicola Valley a little over 24,000 acres. The allotment averaged 450 acres per family, compared with the maximum of 320 acres on a white man's preemption.

Land usually changes hands on a large scale when wars are

fought; in Canada, white dominance settled the redistribution of land. With this final creation of reserves, the Okanagans lost many square miles of hunting range and gained a comparatively large reserve comprising open flat land enough for all to farm. As they had no tradition of farming, this was small recompense. For over 60 years the white fur traders at Kamloops had been farming grain, potatoes and vegetables. Some of the Indians had thus learned by example to plow, seed and harvest.

However, many of the Indians found working for the white settlers more remunerative than farming for themselves, and their land lay untended. The problem came to a head in 1880 when the Indian agent, Albert Elgin Howse, stated that "to the avaricious nature of the chief of this band can be traced the cause of much of the laxity that exists. He entertained the idea that all the land allotted was his, and disposed of it to his favorites only. Thus, only a few have been allowed to cultivate land. The partiality shown by the chief has caused considerable discontent, and in consequence he has lost nearly all his former control."

Then Guichon set out to disprove Sproat's remark that nothing could be cultivated at Chapperon. He built a large wooden frame, set his vegetable seeds in the earth inside, and covered everything carefully each night with a blanket to keep out early frosts. With such care, his garden continued to thrive well into October.

October was also the month when Old Douglas started bringing his cattle in from the fall ranges, but of late he had been journeying to California for the winter, because his tubercular chest was most troublesome during the cold months. During the winter of 1878-79 he went to Santa Barbara and stayed at Mrs. Cross's boardinghouse.

Mrs. Jennie Cross (née Meikle) had been born in New York State and, aged 18, had emigrated with her parents and her brother to New Zealand. On the voyage she fell in love with Captain Cross, the officer in charge of the vessel, but her parents did not approve of a marriage. When the ship landed in New Zealand, the couple ran away together.

Eight months after, Captain Cross was lost at sea and later

Jennie Cross returned with their son to America, settling at Santa Barbara and running a boardinghouse. It was here that she met John Douglas. As it happened, things were not going well for Mrs. Cross, and creditors were demanding payment. Douglas offered to pay all her bills if she would return as his wife to his ranch at Douglas Lake. Douglas was 50, twice her age.

This decision was not to be made lightly, and Jennie Cross wished to know more of her future home. Douglas found an old copy of the *Colonist* in his belongings, and showed it to her. The *Colonist* showed her something of life in British Columbia's capital, Victoria, but of the primitive life at Douglas Lake it told her nothing. The coupled were married on 26 May 1879.

Not long after their marriage, John took his wife and her young son, Willie, by steamship to Victoria. Arriving on a Sunday, Mrs. Douglas wanted them all to attend church service, but Bob Hamilton and Napoleon Sabin had brought down steers from the Interior and among them were some of John's. He went to make arrangements about his beef.

The readiness with which John Douglas had abandoned going to church with his family in favour of looking after his beef disconcerted young Jennie. To appease her, he bought a fine carriage and small piano. They set off by steamer to Yale, the carriage and all, then over the rough wagon road from Yale to Cook's Ferry and the bumpy trail from Nicola to Douglas Lake. The government had spent no money on this route and it was barely fit for a wagon, let alone a carriage.

If she had misgivings on the trail, Jennie was in no way prepared for the earthen-floored, 7-foot-high log cabin that was to be her Douglas Lake home. That night Mrs. John Douglas slept at the McRaes' and the next day refused to stay any longer at Douglas Lake. She wanted to run a boardinghouse in Victoria, so John Douglas packed up the carriage and took his wife and stepson back to the coast, then across to Vancouver Island and the civilization for which Mrs. Douglas yearned.

In Victoria, they bought the appropriately named Douglas House on the corner of Courtenay and Gordon streets, where a beautiful view of James Bay and Victoria Harbour could be had. It was already a private hotel with accommodation for 40.

John Douglas headed home to his cattle ranch. It was just as well that he had ensconced young Jennie at the coast, for the approaching winter would be remembered in British Columbia's history for two devastating reasons—outlaws and weather.

CHAPTER TWO

The stock owner enjoys a very independent life, does not by any means
overwork himself, has plenty of spare time on hand, abundance to eat
and drink, and always appears to have more or less cash in his pocket,
which he spends liberally.

Hugh B. Walkem, *Inland Sentinel,*
7 April 1881

Willie Palmer was at a loss to know where to start looking for his black stallion, missing for three days since 3 December 1879. Telling his Irish wife, Jane, that he was going to search once more, 34-year-old Palmer saddled up and left his homestead north of Nicola Lake. He had settled there amidst the Moore family's holdings in November 1873.

He stopped at Stump Lake on his way north through the snowy landscape and chatted with some of the settlers there. James Kelly, a shepherd for Peter Fraser, who had a sheep ranch at the north end of Stump Lake, had his own problems. Someone had broken into his cabin and wrecked everything.

Continuing north, at the foot of Long Lake Palmer came across four horsemen trailing a pack horse. They were the McLean boys and Alex Hare. Palmer saw his prize stallion, too: 16-year-old Charlie McLean was riding him.

Willie knew this gang well by repute, and Archie, the youngest, personally, for he had hired him to herd cattle back in 1875. The youngest McLean had then been only 11 years old, but, as Palmer said later, "could drive cattle then as well as I could."

Led by Allan McLean, the eldest at 25, the gang were well-known horse stealers, cattle thieves and general hell-raisers. So with four guns pointing at him, Palmer discreetly withdrew after a brief exchange of words.

The McLeans' father, Donald, had led a wild career as a Hudson's Bay chief trader at many posts. When in charge of

Fort Alexandria, he set off with a posse in January 1849 to arrest an Indian, Tlhelh, who had been accused of murder. He did not find Tlhelh but killed his uncle and son-in-law, and a member of the posse wounded a stepdaughter of Tlhelh's and killed her baby. Then McLean bribed another of Tlhelh's uncles, Neztel, to kill his own nephew. McLean also threatened Neztel's life if he refused. Neztel did kill Tlhelh.

McLean too died violently when Anulatlk, a Chilcotin Indian, shot him in the back during the Chilcotin Indian uprising of 1864.

Donald McLean's death left his Indian widow, Sophia, with five half-breed children to raise alone. Of the five, Allan, the eldest, was only ten and the youngest, Archie, was not yet one. A widow's five-year government pension of £100 a year was insufficient. Neither whites nor Indians completely accepted the boys who became as wild as Donald had been. When Charlie McLean had a fight with a Kamloops Indian in 1877 he bit off the Indian's nose, for which he spent three months in jail.

John Edwards, justice of the peace of Kamloops, issued a warrant for the gang's arrest on hearing Palmer's story of his stolen horse. John T. Ussher, the constable, government agent, jailer and court registrar for Kamloops, was to carry it out.

The search for the McLeans and Alex Hare began on Sunday afternoon, 7 December 1879. Ussher and Palmer set off southward from Kamloops. On the way, they met and persuaded Amni Shumway to join in the search. Shumway, a grizzle-bearded Mormon freighter who had taken up a small acreage in 1876, was a good tracker. They spent the night with William Roxborough at the Dominion Government Camp. Roxborough was in charge of the mules and horses of the Canadian Pacific Railway survey crews who were examining possible routes to the Pacific coast.

At seven o'clock on Monday morning, the party set off on the hunt once more, its numbers swelled to five with the addition of William Roxborough and John McLeod, a former Glasgow policeman who had been sheep ranching with his brother above Shumway Lake for 12 months.

As the five horsemen rode south along the old Hudson's Bay Company brigade trail, Roxborough left them to check on some mules he had spotted in the bush. Shortly after, Shumway saw some horses in a large grove of trees. Expecting to find the McLeans' camp, Ussher led the way, calling, "Don't worry boys, they'll never fire a shot."

With that, there was a *crack* and a bullet from Allan McLean's gun just missed its target of Palmer, but pierced McLeod's cheek. McLeod and Palmer retaliated as best they could, though their cap-and-ball shotguns kept misfiring and the gang kept well hidden behind rocks and trees. As McLeod tried to work the gun he had borrowed from a neighbour, he was hit again, this time in the knee. Three bullets lodged in his horse.

Getting out of the direct line of fire, Ussher dismounted, leaving his revolver in the pommel of his saddle, walked up to Alex Hare and tried to talk some sense into him. All too soon, Alex had drawn his knife, the two were grappling and Ussher was down. Then there was a loud report. Smoke rose from the gun held by 15-year-old Archie. Ussher jerked convulsively— and was dead.

Appalled by such ruthlessness and now outnumbered, the special constables retreated while the McLeans waved Ussher's hat mockingly at them. They rode swiftly to Kamloops where a posse was raised, led by Andrew Mara, Ussher's partner in a sawmill at Shuswap, and John Edwards, the Kamloops justice of the peace who had issued the arrest warrant. Twenty armed men set out and took most of Monday returning Ussher's stripped and battered body to Kamloops.

In the meantime, the McLeans and Alex Hare had continued on the rampage. Just a few hours after killing Ussher, the gang came across James Kelly watching his sheep at Napier Lake. He had been an enemy of Allan ever since they had herded together and Kelly had refused to share his food. Two shots, one from Allan and one from Charlie, killed the shepherd.

Sporting Ussher's hat, coat, gloves and boots, and Kelly's mouth organ, small revolver, watch and chain, the McLean brothers and Alex Hare continued south through the Nicola

Valley, stopping at many ranches, stealing more weapons and ammunition, boasting of what they had done, and threatening to chase out the white settlers.

The gang went to Spahomin village to persuade Chief Chillihitzia to lead his people in a rebellion against the whites. Allan was married to one of the chief's daughters and hoped that this relationship would win the chief's help, but Chillihitzia refused to break the peace. The gang stayed the night in a small cabin belonging to Basil, Allan's brother-in-law.

On the Tuesday a sleigh carried Kelly's body to Kamloops and more posses searched the Okanagan and Nicola valleys. The McLeans were relaxing in their safe hide-out at Spahomin, for Allan needed time to recover from a flesh wound he had received during the Ussher battle.

That night, George Caughill, a special constable who had been searching for the McLean Gang for almost three weeks for other offences, paid $100 to a young half-breed informant and learned of their hide-out at Spahomin. He sent two messengers for help, one who rode quickly to the sheep ranch of John Clapperton, justice of the peace at Nicola, then south down the valley to gather more armed volunteers. The other messenger rode north to Kamloops, where John Edwards organized a posse.

Caughill enlisted the help of the Douglas Lake settlers, and by Wednesday morning the local posse had surrounded the small cabin. Many of its members were without arms. In the afternoon, John Clapperton arrived with a posse of 12 men and took control. Chief Chillihitzia and other Okanagans showed their disapproval of the gang by joining the besiegers, so that there were 22 whites and 15 Indians around the shack that night. John Edwards and his Kamloops posse joined the group the next day.

Clapperton sent a message to the McLeans with Johnny Chillihitzia, the chief's son, who came back and said: "The boys say they will not surrender, and so you can burn the house a thousand times."

Clapperton was sure that a siege would starve them out, especially as more men were on the way to help—including an official posse from Victoria under Police Superintendent

Todd. The biggest problem was that they had few guns, while it was known that the gang's cabin was a veritable arsenal.

By now the gang needed water, and Archie, covered by heavy fire, tried to fetch some from the half-frozen Upper Nicola River flowing nearby. The volley of shots from the posse sent him scurrying back to the cabin. Then the whole gang tried to get to their horses to escape, and in the resulting fire two besieging Indians and three of the gang's horses were injured.

The next day, different attempts were made to capture the gang. Pike Richardson roped the cabin chimney, hoping to pull the small building down, but it would not budge and Pike had to drop his rope and spur his horse away as the gang opened fire on him. Posse members repeatedly tried to advance on the cabin under cover of a smouldering hay wagon, but each time were beaten back by the McLeans' fire.

Then a battering ram of logs was built on a wagon, and the contraption so worried the gang that they waved a truce flag and asked a messenger what was happening. They then sent a message which read: "We will surrender if not ironed, and supplied with horses to go to Kamloops." The posse leaders agreed. A volley of shots was fired into the air as the McLeans and Hare, conforming to tradition, spent their remaining ammunition before coming out with their hands held high. The McLeans' last stand was over.

Edwards gave a signal and the posse sprang at the outlaws, holding them down until they were handcuffed. The gang was furious that Edwards had broken his word. The four were stripped of Ussher's and Kelly's clothes, mounted on mules with their feet tied securely under the mules' bellies, and taken to Kamloops. The Douglas Lake and Nicola settlers returned to their hungry stock.

Thirteen months later on 31 January 1881, Alex Hare and the three McLean brothers were hanged at New Westminster.

Now the ranchers in the Nicola Valley were facing another problem: the winter of 1879-80 was turning out to be the worst in memory. Until then, winters had been relatively mild. Certainly snow had fallen, but often Nicola Lake had remained open, and cattle had been able to graze on the tall stands of bunchgrass left unflattened by the snow. As a result, few ranchers put up much winter feed.

The first snows had fallen on 7 November and kept on falling, covering even the tallest stalks of bunchgrass. On New Year's Day 1880, many of the ranchers began feeding their stock from their small hay piles. January progressed and the mercury fell lower, in many places to 40 degrees below. The continuous freezing weather and deep snow demanded more and more feed to keep the cattle, sheep and horses alive. With feed running out, stock began to die.

The settlers had considered the winter of 1871-72 severe, but by 2 March of that year, green grass covered the hillsides, wild ducks were returning and the cattle were getting fat. By 1 March of 1880, many settlers had begun stripping trees and cutting brush for cattle feed. Others killed and burned their starving cattle. In the next two weeks another foot of snow fell, and the sheep ranchers who were trying to keep their flocks alive with grain were desperate. John Douglas, one of the few ranchers who had put up much hay, began selling feed to Peter Fraser, Trapp & McDonald, and John Clapperton. For a month, these ranchers hauled Old Douglas's hay to their thin sheep.

The days were warming up a little to between minus 20 and minus 5 degrees Fahrenheit now, but cattle were dying at the rate of 150 a day, and dead cattle covered the valley.

A quarter of the Nicola's 9,000 head were dead by early April. Any rancher who still had feed left, and there were very few who did, was still feeding. The deep snow, barely starting to recede, hid a multitude of carcasses. One of the sheepmen had lost half his flock. A small rancher who had had 50 head going into the winter had just three calves left. Another had lost over 30 of his spring calf crop, half his expected total. Salvaged hides draped many a fence.

At last the cold began to let up. The warm weather came slowly, lessening the chances of severe flooding. The ranchers could at last get around to inspect the damage. It was a pitiful sight. Many sold out and left the valley, ruined. Others, more optimistic and with some capital left, remained, but swore to put up hay for future winters.

So discouraged was the heavily bearded Thomas Trapp that he dissolved his partnership with Richie McDonald and went to the coast. Clapperton was one of the optimists who stayed.

After the vicious winter of 1879-80, many expected that the next would be mild. It started out that way, encouraging 12 Interior stockmen to drive their cattle over the Coquahalla Pass in December. All through the Christmas season Hugh Murray of Douglas Lake, and A.E. Howse and Richard O'Rourke of Nicola trailed their beef over the treacherously icy route to Hope. Nineteen head slid off the trail, falling to their deaths in the valley below. One steer lagged so far behind that the riders abandoned it to the bears. For the 107 animals delivered to Victoria, they received a pitiful $15 a head.

Spring break-up came early, and by March there was a great demand for plows as well as for the services of Richard O'Rourke, blacksmith. A.E. Howse, who was involving himself more and more in the affairs of Nicola, anticipated the annual shortage of farm equipment. He wrote to the *Victoria Standard* remarking that the farmers of Nicola knew more about Baker & Hamilton, San Francisco, than about any British Columbia farm machinery dealer. He urged agricultural businessmen at the coast to "send up a few mowers, fanning mills, and do not wait until haying time is over. . . . If some man with a little energy would send a lot of plows, harrows, rakes etc. up here, he could be well paid for all the trouble besides establishing a name for himself."

The labours and the attributes of the Nicola Valley were not going unnoticed. Hugh B. Walkem spread its fame far and wide with his articles in the *Ottawa Citizen*. Walkem was the youngest son of the premier of British Columbia, lawyer George Anthony Walkem, and had come to the valley to recuperate from a throat disease.

Walkem's ramblings were as delightfully sugary as cotton candy, and if one were to believe him, the valley was the most perfect place on earth. Praising the superb agricultural qualities of the land, he noted that an acre would yield either 40 bushels of wheat, 60 bushels of oats, or 20 tons of potatoes. Indeed, one of the Mickle brothers had raised seven tons of

potatoes off a quarter acre of bottom land, Bob Hamilton
had grown a 36-pound turnip, and John Gilmore had grown a
turnip weighing 50 pounds. As a fruit growing region, the
Nicola supported "excellently" such fruits as currants,
gooseberries, raspberries. But he did recognize the limitations
in the valley set by the "frosts which are liable to come at any
season of the year," impeding the maturation of pears,
peaches or grapes.

On the subject of stock raising, Walkem's narrative was
positively glowing.

One will see animals whose sides are literally shaking with fat, and
which are better than the best stall fed.... A two year old steer,
killed and dressed will weigh from 700 to 900 lbs, and many will
weigh more. Mr. Douglas, of Douglas Lake, Nicola, killed one
twenty months old that weighed when dressed 743 lbs, and Mr.
Hamilton, a stock-raiser, says that he can go out on the range and
find, without any trouble, three year olds weighing, when dressed,
1,100 lbs; 900 to 1,100 he considers a fair average for three year old
steers.

What he was observing was the tremendous ability of the
Durham or Shorthorn breed of cattle to fatten quickly and
well on grass.

"The largest stock owners in Nicola," Walkem continued,
"are Mr. Guichon, who owns 2,000 head of cattle; Beak,
1,500; R. Hamilton, 1,100; Douglas, 1,100; and Moore
Bros., 1,000."

Turning to horses, he commented that the Guichons, John
Gilmore, Bob Hamilton, the Moore brothers and Mickles all
had imported stock. "These gentlemen deserve a great credit
for their endeavours to introduce a superior class of animal.
The general purpose horse of this region, however, is a
'cayoosh' [cayuse], a small, but hardy native animal." He
went on: "In bucking the animal arches his back, puts his
head between his front legs, stiffens his limbs, springs into the
air and comes down 'all fours', and ... the rider consequently
receives a jar which very often sets all the conflicting emotions
and feelings of the mind considerably on the jar."

With all the produce growing in the valley, greater markets
were needed. The Cariboo market was almost history and the

market at the coast was still difficult to reach, but a far closer market was now readily available, for as a result of Premier Walkem's loud cries for British Columbia to secede if the federal government did not build the long-promised railway connection to the east, Minister of Railways Charles Tupper had called tenders for the Yale-Kamloops section of the CPR. By the spring of 1881, contractor Andrew Onderdonk had 2,000 Chinese workers tunnelling a route through the Fraser Canyon cliffs. Here was the market. By 1882, the little town of Nicola had grown, having acquired new stores, hotels, a community hall, a doctor's office and a religious boarding and day school.

The area prided itself on its agricultural versatility, despite the earlier comment of Sandford Fleming, engineer-in-chief of surveys for the CPR, who stated that "agriculture proper... must always take a secondary place in the interior, and stock raising constitute the chief wealth of the country."

Within a few years Fleming began to be proved correct, and by prominent men who recognized opportunity in the 7,000 white and Chinese labourers working on the CPR's Interior route in 1882. What followed fully utilized the major resource of the Nicola Valley—the native bunchgrass.

CHAPTER THREE

SEALED TENDERS
Will be received at the Office of the
DOMINION GOVERNMENT AGENT,
VICTORIA, B.C.
UP TO SIX O'CLOCK P.M. OF
Thursday, the 16th day of June inst.,
For supplies of the following articles to the
DOMINION GOVERNMENT ENGINEERING PARTY
EMPLOYED ON THE
Canadian Pacific Railway in B.C.
... Fresh Beef per Pound...

H.S. ROEBUCK, Secretary,
Victoria, B.C., June 7th, 1881

Joseph Blackbourne Greaves (pronounced Graves) was born in Pudsey, England, on 18 June 1831. He lost his mother, gained a hated stepmother, attended as little school as could teach him the rudiments of reading, spelling, writing and arithmetic, and became a butcher like his father. At the age of 14 he bet successfully on the horses competing in the Great Yorkshire Stakes, and as he had already run away to go to the races, he decided to leave Yorkshire altogether. He found swineherd's work on a sailing ship, the *Patrick Henry,* which was bound for America, and for 60 days of rough Atlantic seas, he fed and mucked out after pigs.

In the harbour of New York City, Greaves parted with the pigs and over the next nine years travelled over the settled eastern states until in 1854 he caught western fever and travelled by wagon train from Independence, Missouri to California. There he became associated with John B. Fisk and Asa P. Seeley, butchers at Michigan Bar, but by 1864 their goldfields market was dwindling. Hearing of the Cariboo gold rush, Greaves sold his share for cattle and sheep, and drove them north. It was this desire to seek new sales opportunities that was to set him apart from his contemporaries.

He sold those first animals in the Cariboo and returned to Oregon the next year to gather another drive, but this time found no immediate market. He retraced his steps south, turned his cattle out on the shores of the Thompson River, and became a butcher again, spending the first few seasons at Soda Creek on the Fraser River.

Greaves returned to the Williams Creek goldfields in June 1867, and after a fling at butchering there with Isaac Van Volkenburgh, decided to head back to California. His route took him close to the Thompson River; out of curiosity he detoured to locate the herd of cattle he had abandoned there two years earlier. Not only was the original herd still alive but also calves, yearlings and two-year-olds were running with the bunch. He immediately branded all the unmarked cattle—the slickears—and left them once more.

During his second long stay in California, Greaves contracted fever and spent many months in a hospital run by Catholic nuns. This gave him time to ruminate on the direction of his life, and when recuperated he returned to the cattle herd he had twice abandoned in British Columbia's Interior. The original cattle as well as the older offspring were now of marketable age. Greaves decided to stay with his herd on the south bank of the Thompson River below Savona's Ferry at a place later called Walhachin and to raise more cattle for market.

Each year Greaves had more steers and cows to sell from his Thompson River grasslands. The market in the Cariboo was over, but coastal markets were taking its place. Greaves drove his cattle to Cache Creek, then alongside the Fraser River to Yale, and from there took them by steamer to New Westminster and on to Victoria, sometimes taking as long as a month for the journey. In Victoria, the London Market and the Queen's Market became steady buyers of his beef.

Greaves had a great deal in common with John Wilson, his Thompson River neighbour to the west. They were both in their late thirties, from simple folk in northern England, and they delighted in wearing their oldest clothes. Both had gone to the goldfields in California, on the Fraser River, and in the Cariboo. Wilson had found gold at the "Tinker" claim on Williams Creek and, after taking a herd of cattle to the Cariboo, had preempted a quarter section below Savona's Ferry. His land holdings grew with additional purchases along the Thompson River, at Grande Prairie (now Westwold) and later at Cache Creek.

Unlike Wilson, Greaves ranched four years without owning his land at Walhachin. During that time he accumulated capital and was ready to buy land. In 1872, the same year that Old Douglas and his nephew Jack took up land at Douglas Lake, Greaves recorded a preemption for the 160 acres he had staked near Savona's Ferry and for the neighbouring quarter section originally preempted but since abandoned by Stephen Tingley.

The alliance between Greaves and Wilson strengthened in the mid-'70s when Greaves married Mary Ann, the daughter of Wilson's now deceased Lillooet Indian wife. Twenty-year-old Mary Ann, whose natural father, Tom Cavanaugh, had been a gold miner, bore Greaves four children: Joseph Benjamin, Peter, Alice and Mary.

Greaves's ranch grew in 1879 with the purchase of adjoining Crown land. He now owned almost 1,000 acres along the south bank of the Thompson, but the next year he was looking much farther afield, to the market consisting of the work gangs who were building the Interior section of the Canadian Pacific Railway.

In June 1881, the CPR invited anyone who could guarantee a large and steady supply of fresh beef to send a sealed tender to Victoria. Greaves realized that such a large work force would require an enormous number of cattle, improving beef prices throughout the province. He knew he had little chance of maintaining a steady supply on his own, and wondered whether he could find enough financial backing to be able to acquire sufficient cattle for the CPR contract.

Among Greaves's many acquaintances was someone who understood the potential of such a market, and that was Benjamin Van Volkenburgh, an experienced butcher from the Cariboo mines market.

There were three Van Volkenburgh brothers: Benjamin, Abraham and Isaac, Americans of Dutch extraction. Together with Jerome Harper and Edward Toomey, the three brothers formed a company, Messrs. Van Volkenburgh & Co. To the firm, Toomey brought his butchering experience and his Barkerville store and abattoir; Harper brought his experience in buying, driving and selling cattle, and the assurance of a good supply of beef from the ranges he was beginning to acquire; Benjamin, Abraham and Isaac brought the manpower with which it was possible to open butcher stores in Barkerville, Richfield, Van Winkle and Grouse Creek.

This butchering company was the first of its type in the Cariboo. It appreciated the need for a steady winter supply and quickly grasped a virtual monopoly on the beef trade within the mining community, a monopoly which it held for more than a decade. Fresh meat became synonymous with the name of Messrs. Van Volkenburgh & Co.

In 1865 the Cariboo depression began—a slump caused by the need for steam machinery to mine ever deeper, and the resultant exodus of those without the funds to acquire such expensive equipment. Bigger mining companies employing fewer men held the majority of claims. One commodity affected was beef. Until 1865 it had sold for 35 cents a pound; the depression dropped this price by a dime. Hides fell to 10 cents a pound, a value that did not cover the cost of freighting them to the coast.

When gold was found in faraway Omineca in 1869, miners in the Cariboo rushed to the new strike, decreasing the Cariboo meat trade still further. Not long after the entry of British Columbia into Canada in 1871, the manager of the Victoria branch of the Bank of British Columbia, William Curtis Ward, visited the goldfields. His findings, reported back to London, reflected the continuing decline.

Nevertheless there were still miners in the area to sustain Messrs. Van Volkenburgh & Co.'s chain of butcher stores for a while. The annual yield of gold still hit the million-dollar figure, and the population, in 1874 comprised of 920 whites, 685 Chinese, 570 natives, and 32 coloured persons, still consumed beef—and liquor, as evidenced by the ten saloons in Van Winkle.

This diminished market during the '70s, the death of Jerome Harper in 1874, plus the increasing business opportunities elsewhere in the province, all caused the Van Volkenburgh brothers to leave the Cariboo. For over a decade they had run a profitable business and made a great name for themselves. Now they branched out as individuals. Abraham moved his wife and daughter to Yale, where he opened a butcher store on Front Street. Isaac began ranching on Canoe Creek, near Clinton. Benjamin Van Volkenburgh moved his wife and three children to Victoria.

In February 1880 Benjamin purchased for $11,250 the unexpired portion of a lease on the Alhambra Building from Charles W.R. Thomson, secretary of the Victoria Gas Company. This centrally located structure on the corner of Yates and Government streets had been the Brown Jug Saloon. Van Volkenburgh altered its interior and opened it as the British Columbia Meat Market that spring. Thaddeus Harper, Jerome's brother and successor to all his property, agreed to send down his marketable cattle. Van Volkenburgh built a wharf in Cadboro Bay to unload Harper's cattle, sheep and pigs. He fenced in grazing for the stock prior to slaughter and built an abattoir close by.

His biggest competitor was the Queen's Market, which Lawrence Goodacre had run for the past 11 years, and though Goodacre specialized in supplying the many shipping companies, he had cornered a large number of the other customers.

These two men—Van Volkenburgh, whose years in down-to-earth Cariboo had entrenched his straight-laced style, and Goodacre, who enjoyed displaying his wares lavishly— became great friends privately, though they harangued each other's stores through the pages of the *Colonist*. It was light-hearted bantering and an excellent form of advertising, very necessary considering that the butcher stores of Victoria stocked as much as 100,000 pounds of meat for the consumption of the 6,000 inhabitants.

But Van Volkenburgh's thoughts were elsewhere. He had received a letter from Greaves.

Cache Creek Decr 24 1881

B. Van Volkenburgh
Dear Sir

Enclosed please find statement of cattle. The number included cattle for next season's market. The number is larger than I thought it was but I think next season will require most of them. If we go into the market we will have to purchase four thousand head to affect anything. I think the steer cattle can be bought for 20 or 21 dollars per head. Four thousand head of steers will control the market next season and there is many in it. If the weather keeps mild there will be some fat cows early in the spring that will have to be bought to keep them out of the market. If we take advantage of the spring market we will have to buy at once. The weather is so fine the cattle will keep fat all winter without being fed and everybody will be driving early in the spring.

If we don't go into the market I think cattle will be low this spring and we might buy cheaper in May than now, but we would lose this spring trade.

I mention this so you can think over it.

Any arrangement you make I am ready to go to work any time.

We are having the finest weather I ever saw for this time [of] year. No cattle die this winter.

Truly yours
J.B. Greaves

Greaves's idea was that for $80,000, sufficient cattle could be bought to control the cattle market in British Columbia and also to secure the contract to provide beef to the CPR workgangs. This was a bigger sum than the two men could raise. It was now up to Van Volkenburgh to interest other Victoria businessmen in supplying the rest of the financing.

When the Hudson's Bay Company received a grant from the British Crown in 1849 for Vancouver Island, it undertook to bring out colonial settlers within five years and so needed a surveyor to map the land available for colonization. In June 1851, Joseph Despard Pemberton, of cool demeanour and with heavy sideburns and wavy moustache, arrived in Fort Victoria.

Born in Dublin, Ireland, 30 years earlier, grandson of one of the Lord Mayors of Dublin, Professor Pemberton had already designed four British railway lines and taught surveying, civil engineering and mathematics at the Royal Agricultural College, Cirencester, England.

In his first year, the zealous Irishman surveyed the island to the southeast, surveyed the 6-square-mile fur trade reserve, planned townsites for Victoria and Esquimalt, and became the first white man to reach Cowichan Lake. Later, he was part of the successful coal-finding exploration of Wentuhuysen Inlet, a place he retitled with its Indian name, Nanaimo. He undertook a trigonometrical survey of the island from Sooke to Nanaimo; London cartographers printed this first reliable map of Vancouver Island. In 1855, he purchased "The Gonzalo Farm" at Oak Bay, where he eventually owned 1,200 acres. Pemberton entered politics in August 1856, sitting in the first House of Assembly and serving Victoria for 7 out of the next 12 years.

As first colonial surveyor for the mainland colony of British Columbia, Pemberton laid out townsites at Fort Langley, Fort Hope and Fort Yale, and in July 1860 became surveyor general of Vancouver Island. He travelled to Great Britain twice in the next four years, for the publication of his book *Facts and Figures Relating to Vancouver Island and British Columbia* and to purchase machinery to improve Victoria's Harbour, bringing home with him his new bride, the aristocratic Teresa Jane Grautoff.

In October 1864, shortly after a bad riding accident, Pemberton resigned from all duties. An excellent judge of horseflesh and passionately fond of anything equine, Pemberton imported some of the first Clydesdales and Percherons into British Columbia. As well, he began raising his own herd of Durham cattle.

Just to the north of Pemberton's Gonzales estate was Benjamin Van Volkenburgh's wharf. For 22 months these two had been neighbours, and Van Volkenburgh persuaded Pemberton to join the venture of supplying meat to the CPR workmen.

By the age of 21, William Curtis Ward, the well-educated son of a Winchester milkman, had decided to make a career of banking. Having some experience as a cashier in the National Provincial Bank in Bristol, he obtained an interview in December 1863 with the London Court of the Bank of British Columbia. The directors liked the strong-featured, bearded young man, and hired him as an accountant for their branch in Victoria, British Columbia. Ward's starting salary was $1,500 per annum, his probationary period five years.

Before departing Ward married Lydia Sothcott, a comely 17-year-old English girl. They arrived at Vancouver Island after three months' sailing, and made their home in rooms above the bank.

The Bank of British Columbia, in competition with Wells, Fargo and Company's Express, Barnard's Express, Macdonald's Bank and the Bank of British North America, launched into an expansion program—for there was a surplus of gold dust in the young colonies, a lack of anywhere to deposit or sell it, and a lack of coinage—and soon opened branches at New Westminster, the Cariboo, San Francisco, Yale, Quesnel Mouth and Nanaimo. Unused to the unstable economy of the colonies—based as it was on gold mining—the first two Victoria managers supervised the other branches very loosely, allowing vast overdrafts to accumulate. In March 1867 at the age of 25, Ward became the third manager in Victoria, taking complete control over all the British Columbia branches.

Even though employment for some miners was on a more permanent footing in the Cariboo through the '70s, many turned to other occupations, some working on the Fraser River fisheries, others in the lumber mills at Burrard Inlet, still others becoming farmers and ranchers in the Fraser Valley and the Interior.

Ward's "well established business perspecuity" and the

healthier state of the young province brought the bank's balance sheet from an 1865 overdraft of $1 million to a surplus of £1 million in 1875. The young banker had become one of the most important financial figures in British Columbia. His capabilities deservedly brought in an excellent salary of $4,000 annually, and the bank built him spacious ''Highwood,'' a beautiful brick home set in ten acres of formal gardens, orchards and tennis courts. Ward gave his ten children the best of everything.

In late 1881, Benjamin Van Volkenburgh approached Ward to suggest he invest in the embryonic cattle venture. Ward decided to discuss the plan with two of his business associates in Victoria.

When Charles William Ringler Thomson came to the west coast to seek his fortune in late 1859, he had no intention of mining. Instead he made a study of the gold rush, observing that the miners were travelling farther into the Interior, where provisions were costly because of the distances travelled and the lack of pack animals. He had noticed an excess of mules on the Sandwich Islands when strong winds blew his ship off course there, so the 32-year-old son of a London solicitor sent for a boatload of mules, which he sold at Yale for a profit. He was also profitably involved in commercial fishing.

Spending part of his first winter in mild California, Thomson met Matilda Mills Midwinter, a widow four years his senior, and married her in San Francisco on 15 March 1860. Returning to Vancouver Island, Thomson and his wife outfitted a barge with provisions, purchased some cattle, and journeyed to the Saanich Peninsula. They were among the earliest farmers in that most arable region.

That fall, Thomson invested in the Victoria Gas Company and two years later became the company secretary. By the winter of 1862-63, the company had supplied Victoria with gas lighting. As Thomson's knowledge of the business improved, so did his salary; from $100 a month in July 1862 to $250 a month in 1867. He became manager as well as secretary.

Now in their fifties, Thomson and his wife moved to a delightful residence in Esquimalt Harbour called "Maplebank." The 20-year-old country seat had been the successive home of two admirals, one naval captain, and the Honourable Joseph Needham. The spacious grounds of Maplebank became home also to a flock of ring-necked pheasants that Thomson imported from England, the first in British Columbia.

William Curtis Ward approached Thomson about the cattle venture. Though Thomson did not know J.B. Greaves, he did know the others, especially Van Volkenburgh, who had purchased from him the unexpired lease of the Alhambra Building for his butcher store. The idea attracted Thomson and he agreed to join in.

The new office that Governor James Douglas created in 1859 well suited Peter O'Reilly, a tall Irishman who had been born at Ballybeg House, Kells, County Meath, 31 years earlier and had been a lieutenant in the revenue police. O'Reilly became one of the six colonial gold commissioners. With Fort Langley as his first post, his duties were many: stipendiary magistrate, justice of the peace responsible for settling all minor mining and civil disputes; collector of miners' licences; recorder of mining claims; assistant commissioner of lands; revenue collector; Indian agent, and coroner.

As the throng of gold miners in British Columbia advanced from one goldfield to another, so did Gold Commissioner O'Reilly. During his years in the Cariboo, O'Reilly met Joseph W. Trutch, contractor for a section of the Cariboo Road, and his youngest sister, Caroline Agnes. Carry Trutch and Peter O'Reilly married on 15 December 1863 at Christ Church Cathedral, Victoria.

Peter O'Reilly joined the first Legislative Council for mainland British Columbia in January 1864 as one of its five magistrates. He stayed on the council eight years, two of them with Joseph Despard Pemberton. By the time Victoria became the capital of the united colony of British Columbia, the O'Reillys had moved to that city and purchased a house at Point Ellice for $2,500. It overlooked the narrows between Victoria's Inner Harbour and Selkirk Water.

Becoming a county court judge in 1868, O'Reilly established a second home at Yale and held courts throughout the Interior. These travels took him through the Nicola Valley just as the first preemptors were taking up land. Conferring with his brother-in-law, Joseph Trutch, O'Reilly laid out reserves for both the Okanagan and Thompson Indians living there.

O'Reilly and his investment partners—William C. Ward, who was his bank manager, and Charles W.R. Thomson, who ran the Gas Works of which he was a director—met several times in late 1881 to discuss mortgage transactions. They agreed to pass up mortgaging Thaddeus Harper's property and decided to mortgage Alexander Coutlie's Nicola Valley ranch for $4,000.

Thus it was only natural that Peter O'Reilly was the other person whom Ward approached in January 1882 concerning the cattle venture. "Long chat with Thomson & Ward in reference to the Cattle spec," noted O'Reilly in his ledger under the date 14 January, "when it was agreed that we three should go in with Van V., J.G., & J.D.P. at $5,000 each."

The "Cattle spec" was finally underway.

CHAPTER FOUR

Corner in Beef: It is reported that some shrewd speculators have secured a corner in Mainland beef, and that in consequence meat will have an upward tendency. It is estimated that a cool $150,000 will be made out of the unfortunate consumers, unless the 'corner is broke' by some means not at present discovered.

Nanaimo Free Press *per* British Columbian,
19 April 1882

The dashing yet aloof Andrew Onderdonk had employed 5,000 men during 1881 in the construction of the Yale-Kamloops section of the Canadian Pacific Railway. There would be just as many men the following year, and fresh beef would be needed to feed them. This was the opportunity that caused Joseph B. Greaves and Benjamin Van Volkenburgh to search out Hon. Joseph D. Pemberton, William C. Ward, Charles W.R. Thompson and Judge Peter O'Reilly as financiers to establish a beef price monopoly within British Columbia. If the syndicate were able to buy up all or nearly all the steers and cows ready for the 1882 market, then the success of its tender for the CPR beef contract did not matter to its members. They would control the price anyway, for only they could supply the meat.

In a letter to Greaves on 23 January 1882, W.C. Ward enclosed a memorandum outlining the terms by which the syndicate agreed to operate. It named just J.B. Greaves to represent the $20,000, which Ward "considered best for many reasons."

As syndicate manager in charge of buying and selling the cattle, Greaves would receive $150 per month in addition to his own and the cattle's travelling expenses. He was either to purchase the cattle with a cash deposit and arrange favourable terms for the balance or to secure a very good bargain with an immediate cash payment in full. As the syndicate owned no land, the members would pasture their cattle on the sellers' ranches until needed. Greaves was to keep a thorough record of all transactions, and was to render a full account of affairs whenever the other contributors requested. From time to time he was to send copies of the cattle bills of sale by registered mail to Victoria to keep the other five informed.

Though it was going to take the purchase of up to 4,000 head to corner the beef market, the extensive use of notes of credit plus small cash deposits was going to keep each contributor's subscription down considerably from his share of an $80,000 investment. By staggering the cattle payments, they would not even have to pay in their full $5,000 shares. All cash advances, however, would incur interest at 1 per cent per month, and a majority wish could terminate further advances and even the entire business at any time, whereupon the contributors would assume the profit or loss in proportion to the amounts each had so far advanced. Greaves and Van Volkenburgh could provide their $5,000 in the form of notes bearing interest or could pay with cattle, which is what Greaves did, putting in 233 of his own steers at $21 each.

The speculators of the syndicate with Ward as their spokesman realized that the success or failure of the undertaking depended entirely on Greaves's integrity, honour and good judgement. Nervous about the magnitude of their total possible investment, he warned Greaves, "Nothing but the expectation of a *good* profit would induce anybody to take it in hand in any shape. . . . If you concluded to go ahead, you can draw on the Bank as you need, & Mr. Thomson will pay in to meet cheques as they come forward." Thomson was to be secretary-treasurer for the syndicate.

Secrecy was Ward's biggest wish in the whole affair and he ended his letter to Greaves saying, "every precaution will be observed here as to keeping these arrangements from being known and . . . you will bear in mind that I am wishful not to be known in it in any way. Trusting that you will accomplish all success, & relying upon your bearing me out most thoroughly in the confidence that is placed in you. . . . "

Greaves was so cognizant of the need for secrecy that when it came time to make out a bill of sale for his cattle to the syndicate as his subscription, he preferred not to do it in the Interior "for fear people here would find out if I got the Bill of Sale Recorded what was going on. I thought it best to have it done in Victoria when I come down."

Immediately, Greaves began buying up local cattle, leaving them on the seller's ranch until needed and leaving a note to cover the cost. By 22 February, he wrote: "Friend Ben, I have bought (22) twenty two Hundred Head of the Best Cattle there is in the country. The Nicola Cattle Cost from (20) to (23) dollars per Head. Cattle that is left cannot be bought for less than ($25). I did not think it advisable to pay at that price at present. Kamloops Cattle Cost from twenty to twenty two except Victors. He wants (25). I will take them."

Greaves was concerned that matters could backfire, and instructed Van Volkenburgh, "Now Ben be sure and arrange for the checks that I have drawn on the Bank to be paid. If they should be returned it would break up the whole arrangement, for most of them that I have bought from would go back on the bargain if they had any chance to get out of it. Cattle men are getting excited. They think there is something up."

That same February day Greaves left for the Okanagan and Similkameen country in order to "get (30) or (35) Hundred Head of Cattle that will give our Compy control of the market for this season." Greaves had already sent Brock McQueen, one of the Overlanders of 1862 and now a rancher near Kamloops, to buy 400 cattle on his behalf at the Okanagan Mission at between $17 and $20 each.

Greaves had heard that Thaddeus Harper was also eager for the CPR beef contract and was heading towards the Similkameen country at the end of February to buy cattle. Determined to outwit Harper, Greaves told Ben, "I will try to be there first," and, as an afterthought, "I think we will make some money in this business."

The possibility of their making money seemed high, but first they had to pay some out for the cattle and then receive some back from the butchers. As Greaves and Harper competed over the Okanagan and Similkameen stock, a letter from Ward arrived at Greaves's address at the Cache Creek post office. It showed that the Yorkshireman would have to be more careful in his choice of outlets. "Mr. Robinson of Van[couver] has not yet settled his overdue note to you per $375. but has asked for six weeks further time, & has given Mr. Kiddy's name as endorser for the extension. He expects to buy some more Cattle from you when you come down but until he has paid up your present notes against him, I should think you had better keep clear of him."

The Yale paper, *Inland Sentinel*, had noticed the excitement among beef raisers by 23 March. "Since Mr. Greaves has gone extensively into the purchase of cattle, Messrs. Harper and Van Volkenburgh have also advanced and prices are now ranging pretty high. Stock raisers are in good humour at present." After all, Ward had written from Victoria telling Greaves, "I think that you will do well to continue to buy any desirable lots that you can pick up & even if you have to raise your figure a few dollars per head, you should not hesitate to do so. From all appearances you will be able with the exercise of good judgement to control prices here."

By April, readers of the *Nanaimo Free Press* and the *British Columbian* heard of the syndicate's attempted monopoly. "Corner in Beef: It is reported that some shrewd speculators have secured a corner in Mainland beef, and that in consequence meat will have an upward tendency. It is estimated that a cool $150,000 will be made out of the unfortunate consumers, unless the 'corner is broke' by some means not at present discovered." The newspapers had obviously forgotten the long periods during the '70s when because of the lack of a market, cattle had brought meagre prices and many had just grown old on the ranches that had raised them. Now that the ranchers had begun to earn a fair return for their stock, it was time to pity "the unfortunate consumers."

Towards the end of July the *British Columbian* printed a surprising article which read, "We understand the 'corner' in beef has been broken up and stock men beyond the Cascades are 'as was'"—surprising, because Greaves was still purchasing cattle, and the syndicate was in sound shape. The only change in the marketplace was that Thaddeus Harper had received the CPR beef contract and had begun filling it, in spite of Greaves and the syndicate. Harper was taking the CPR contract just as seriously as the syndicate was taking its virtual control on price, for not long before, Harper had imported seven purebred bulls to increase the performance of his own cattle over the coming years.

Greaves continued to purchase bunches of cattle in various places, and continued to hold them on the sellers' ranches until the coast market required them. Then he herded them towards Yale and shipped them to the coast.

During 1883 the syndicate kept its policy the same and in fact did better than in 1882. Thaddeus Harper again contracted to supply Andrew Onderdonk with beef for his construction gangs and lived up to this contract as best he could. But with Greaves still roaming the Interior on the lookout for any and all likely buys, Harper defaulted in supplying a sufficient quantity, and on behalf of the syndicate Greaves received the CPR contract for beef.

Angus McInnis, who had left his Ontario home in 1876, became Greaves's foreman in charge of delivering cattle to the CPR. Each week he and his riders drove 300 head to the point of construction, sometimes travelling via the Coquahalla Pass, sometimes via the Fraser Canyon itself, depending on the location of the ranches that were supplying that particular drive. Many cattle died in the drives—falling off the rugged cliff paths—but Greaves fulfilled his contract.

The syndicate's expedient cattle purchases that had caused Harper to default improved the low market prices for many ranchers. Greaves was becoming well known through his buying trips on horseback throughout the Interior. Okanagan, Nicola, Thompson Valley and Cariboo ranchers began to put their trust in the persuasive, square-dealing beef speculator who knew stock and who offered fair prices. The notes the Yorkshireman left behind were as good as cash. Greaves had become a timely lending service. Ranchers were taking advances on their cattle before making delivery in the

49

same way that Victoria financiers were writing cheques, with W.C. Ward's approval, on an overdraft.

Towards the end of this the syndicate's second season, Joseph B. Greaves entered into a partnership with a Nicola Valley landowner, Charles Miles Beak. This alliance was to harvest a bountiful crop.

CHAPTER FIVE

I knew Beak well, an extra clever trader, but on the square.

James B. Leighton to Lawrence P. Guichon,
14 October 1938

At 18, Charles Miles Beak of Purton, Wiltshire, went to sea, first as a cook aboard the sailing ship *Hamilton*, leaving Cardiff for the Panama. Between 1855 and 1860 he spent much of his shore time in the port of San Francisco, joining the activities of its Vigilance Committee. As well, he mined gold from the Sacramento River.

In 1862, with two others, he drove 300 head of cattle north to British Columbia, selling the beef in and en route to the Cariboo. Then from Oregon the young Englishman partnered James Dole to drive the first flock of sheep into British Columbia and to open a butcher store in Barkerville; their mutton sold for 40 cents a pound and candles made from the sheep's tallow brought in 50 cents a pound.

Beak married 16-year-old Maria Johnson of Glencoe, Oregon, on 16 May 1868, returned to the Cariboo, and bought the 105 Mile property to start a 100-cow dairy. Beak's dairy shipped butter by Barnard's Express to the Cariboo, where it sold for 50 cents a pound.

Beak soon left his hired men operating the dairy and made another attempt at butchering in Barkerville, ranging his 100 cattle and 500 sheep on Bald Mountain overlooking Williams Creek. He advertised in the *Cariboo Sentinel*: "Cariboo Market: C.M. Beak: Desires to inform the inhabitants of Barkerville, Richfield and vicinity that he has opened a shop in Nott's new building, where he will sell butcher's meat of the choicest description, at From 10 to 15 cents per pound and he trusts to receive a share of business and to 'Live and Let Live!'" Coyotes and bears so troubled his stock, however, that by 1870 he had closed his butcher store once more. In the early '70s Beak began raising beef at Lac La Hache; he also purchased 127 Mile House on Cariboo Road and had the previous owner, D. Pratt, run it for him.

His stay at Lac La Hache taught him that heavy summer and winter grazing had nearly exterminated the native grasses along the Cariboo Road and so he sold his properties there and looked throughout the Interior to find stands of bunchgrass still close to their native state. By July 1878, Beak, his wife and four sons were well settled on land that Beak had preempted on the Upper Nicola River, just east of where Lauder Creek joins the bigger tributary. Beak built for his family a cabin that was to become a landmark in the area.

Directly north of Beak's claim was another 320 acres preempted by Walter Pound Trounce, previously a freighter in the employ of the Hudson's Bay Company. North of Trounce's was land that Lauder had taken up in 1876, just two years before. That summer, Beak sold Joseph Lauder 16 of the dairy cows he had brought with him from the 105 Mile dairy. All these cows had calves at foot. To begin with, the Lauders had no use for the vast quantity of milk that their dairy herd yielded, but when a German traveller staying at the ranch was nursed through an illness by Joseph Lauder's wife, in gratitude the visitor showed his hosts how to make cheese from the dairy cows' milk. This was a luxury item in the Interior, and Lauder was soon receiving a dollar a pound for his cheese.

In his new location, Beak went seriously into the business of buying and raising dairy cattle, beef cattle and pigs. Vast acres of pea vines flourished and soon the pigs were growing fat. His cattle grew in quantity and quality on the bunchgrass. Joseph Castillion milked the dairy herd.

Beak became involved in the McLean Gang outrage in December 1879, when the gang, then on their way to Spahomin, stole weapons and ammunition from his home.

Though Beak was a shrewd speculator and amassed 2,000 head of cattle by the time the ghastly winter of 1879-80 rolled around, he had not made any realistic plans to winter feed them. By the end of January, he was almost without feed for his vast herd and deaths cut it down to 1,500 head. Unlike other unfortunate settlers of the Nicola Valley who sold out, Beak pulled in his belt and continued to increase his herd. He was helped by the thousands of Chinese who entered British Columbia in 1881 to construct the CPR line through the Fraser Valley, for they became a ready market for his pigs.

Beak had registered the brand $C\!\!\beta$ for his cattle on 28 October 1878, but found that it blotched badly and became hard to read. In September 1882, he took another, $7/$, and the same year on 23 November, a third, $///$. This last he kept, and within a few short years it was to become known as the "hundred and eleven" brand.

Beak then left someone to watch the ranch, and with his family left for England, where he persuaded friends into lending him sufficient funds to realize his dream of raising thousands of animals on the bunchgrass hillsides of the Nicola Valley.

In the village of Purton was the workhouse for the Cricklade & Wootton Bassett Union, and Beak signed an agreement to take three of the Union's charges. "Ernest Selman aged 12 years, Frederick King aged 12 years and William Strong aged 11 years" were "to serve him and with him to work and live at Nicola Valley . . . for the term of four years now next ensuing." In return for this work, Beak was immediately to clothe and outfit them "to the value of Five Pounds . . . each"; to keep them equipped in "proper clothing and linen"; to "maintain them with proper food and nourishment and provide proper lodging"; and "at the expiration of the said term [to] pay to each of them the sum of Sixty pounds for his services to be rendered."

Leaving his wife and family in Purton, Beak returned with his three young hired hands to his home in the Nicola Valley. There Beak found matters had not gone as well as expected. Alfred Morton, a new settler from Ireland, had jumped Beak's land claim on a technicality and was now living there himself. Beak had much bigger plans on his mind than one district lot composed of 343 acres, however, and left Morton alone.

Beak started his purchasing campaign on 6 June 1883, by buying John Douglas, Sr.'s 1,400 head of cattle for $36,000 and his 900 acres of land at the head of Douglas Lake for another $4,000. The new owner and his three hired hands moved into the cabin Douglas had built. Next he bought Napoleon Sabin's 400 head and 320-acre preemption on a flat bordering the south bank of the Nicola River. Sabin had received a certificate of improvement for this land in 1881 and all Beak had to do to receive title was pay a dollar per acre. Then he purchased the land of his former neighbour, Walter Pound Trounce.

In September, Beak bought Hugh Murray and Ronald McRae's cattle and land west of Chapperon Lake on the creek. This comprised 312 acres, for which he paid $500. Hugh Murray left for Victoria, but Ronald McRae stayed to prove up (establish title) and buy more land. On 5 October, Beak purchased, directly from the Crown, one district lot comprising 5,963 acres. This land bordered Douglas Lake on its east shore, and with Old Douglas and Sabin's land on the north, Beak now possessed a solid block around the north and east end of Douglas Lake.

At the beginning of October, Beak finally bought Laurent Guichon's land on the northeast shore of Chapperon Lake: almost 500 acres, most of which had first been François Chapperon's Crown land. Beak reportedly paid $44,000 for the land, cattle and horses, the cattle herd having numbered 2,000 in 1880. On leaving the Nicola Valley, Laurent Guichon bade farewell to his brother Joseph, who had moved to the mouth of the Nicola River, and went to the coast, where he had previously purchased almost the same acreage of land in Surrey at what became known as Port Guichon. Charles Miles Beak now owned over 8,000 acres of land, 5,000 head of cattle, and 60 horses in the Upper Nicola Valley. In less than a year he had become one of the largest ranchers in the Interior.

Beak knew the syndicate of men who controlled the price of beef in the province and he sought its manager—J. B. Greaves. Once the two powerful cattlemen had reached an understanding, Beak and Greaves sat down with Montague W. Tyrwhitt Drake, an eminent Victoria lawyer, who drew up the necessary documents.

On 12 October 1883, Beak gave Greaves an undivided half

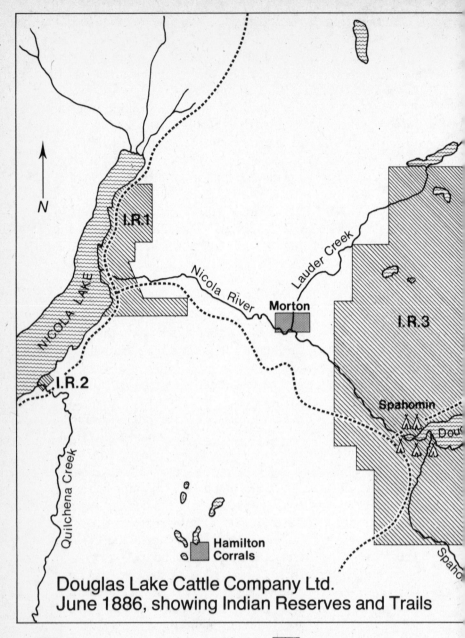

N

NICOLA LAKE

I.R.1

I.R.2

Quilchena Creek

Nicola River

Lauder Creek

Morton

I.R.3

Spahomin

Dou

Spaho

Hamilton Corrals

Douglas Lake Cattle Company Ltd.
June 1886, showing Indian Reserves and Trails

▨ D.L.C.C. land base 4 June 1884	▨ Indian Reserve (I.R.)
▨ D.L.C.C. land base 30 June 1886	⋯⋯ Wagon Trails

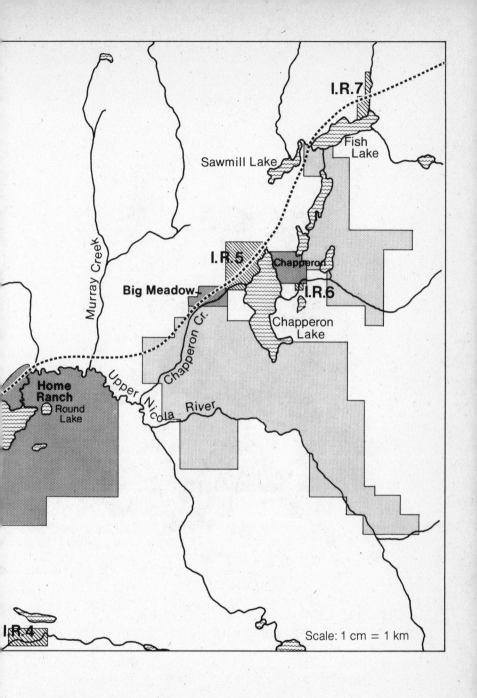

I.R.7

Sawmill Lake

Fish Lake

I.R.5

Chapperon

Big Meadow

I.R.6

Chapperon Cr.

Chapperon Lake

Murray Creek

Home Ranch

Round Lake

Upper Nicola River

I.R.4

Scale: 1 cm = 1 km

John Douglas (courtesy of Provincial Archives of B.C.)

J. D. Pemberton (courtesy of
Provincial Archives of B.C.)

W. C. Ward (courtesy of Provincial
Archives of B.C.)

C. M. Beak, circa 1866 (courtesy of
Humphrey Beak)

Peter O'Reilly (courtesy of
Provincial Archives of B.C.)

Highwood (courtesy of Provincial Archives of B.C.)

C. M. Beak (extreme left) with Indians, circa 1882 (courtesy of Humphrey Beak)

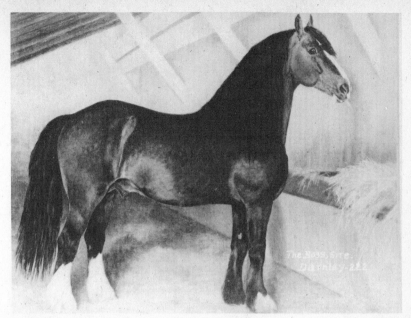

The Boss (courtesy of Douglas Lake Cattle Co. Ltd.)

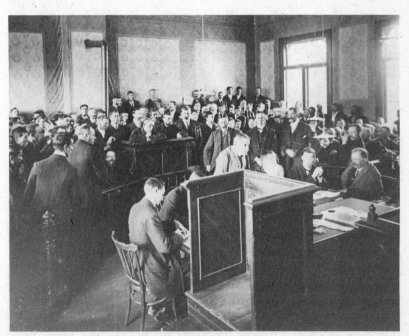

Miner, Dunn and Colquhoun on trial (courtesy of Kamloops Museum Assoc.)

Original Ranch House, Home Ranch (courtesy of Brian K. de P. Chance)

Judging Shorthorn bulls at early Kamloops bull sale (courtesy of Brian K. de P. Chance)

J. B. Greaves (courtesy of E. Neil Woolliams)

Francis B. Ward (courtesy of Betty Farrow)

Kenny Ward, 1917 (courtesy of Curtis
Ward)

". . . being mostly Indians and breeds." (courtesy of Brian K. de P. Chance)

Home Ranch stables and store, 1917 (courtesy of Curtis Ward)

interest in 152 acres of Guichon's land and in Murray and McRae's 312 acres, plus their cattle, horses, crops and agricultural implements, in return for $21,000, due on 1 October 1884, with interest at 7 per cent per annum. The next day, Beak gave Greaves a second agreement of sale for an undivided half interest in Douglas's, Sabin's and Trounce's 1,519 acres, the 5,000 head of cattle, 60 horses, and further hay and equipment in return for $40,000.

These memoranda were subject to an indenture stating that Beak and Greaves were entering into a five-year partnership as "farmers and cattle-raisers." Into the partnership Greaves was to put 1,000 of his own cattle, valued at $22 each. Along with the other financial transactions, this transfer meant that of the original $61,000 owed to Beak, Greaves now owed a net sum of only $28,250. Ten thousand was due in one year, the rest as he could manage it.

The partnership began immediately, under the title of "Beak and Greaves." Beak was to be interim manager at $150 a month. Greaves was free to continue as manager of the syndicate. Another clause allowed Greaves to purchase from the partnership "all steers from two years old and upwards at ($30) thirty dollars a head" on behalf of the syndicate.

From these two groups, the partnership of Beak and Greaves and the syndicate of Greaves, Van Volkenburgh, Ward, Pemberton, Thomson and O'Reilly, was to emerge the final shape of a great cattle business in the Upper Nicola Valley the following year in 1884.

CHAPTER SIX

A Cattle Train: Wednesday at 1:30 p.m. the first cattle train passed through Yale, constituted of five car loads, in all 85 head, and drawn by No. 4 engine. Flat cars had been boarded up at the sides and ends and the cattle appeared to be very quiet. Mr. J.B. Greaves has the honor of shipping the first head of cattle by the Railway down the line. We understand the charge for taking the cattle from the place of loading to Port Moody is $6 per head. This will pay very well and, no doubt, the Railway contractors will follow it up as a business. Loaded with steel rails, etc. up and the cattle back will do.

Inland Sentinel,
14 February 1884

Joseph Blackbourne Greaves was the linking force between his partnership with Beak and the cattle syndicate, managing the beef syndicate and taking an active interest in his cattle-raising partnership with Beak. As did that of many another overly dedicated cattleman, Greaves's marriage suffered through lack of attention, and Mary Ann and his four young children left him. As well, his Thompson River ranch was showing signs of absentee ownership. Early in January 1888, he would sell his entire Walhachin holdings, over 1,200 acres, to his friend and neighbour, "Cattle King" John Wilson.

Charlie Beak assumed the management of the partnership, caring for the 5,000 head of cattle and 60 head of horses, fencing the 8,000 acres of land, and searching out new acreages to acquire.

While activities on the Kamloops-to-Port Moody section of the CPR slowed down for the winter of 1883-84, and although Onderdonk had underestimated the cost of laying track through the Fraser River Canyon and was in financial straits, the ranchers of the Interior were profiting from the lucrative beef market. On 11 December, Greaves was able to extend $250 worth of credit to Jack Douglas in exchange for "cattle which I agree to Deliver in 1884 at the market rates." The next day on his own behalf Greaves bought John Hamilton's "Mountain Ranch formerly known as the Horse Carralls situated six miles above the Hamilton Ranch towards Minnie Lake."

On 13 February 1884, Greaves made history when he shipped 85 head of cattle from Yale to Port Moody, for these were the first cattle to travel by rail in British Columbia. As it happened, this was also one of the last shipments of syndicate cattle, for Greaves journeyed to Victoria to discuss the future of the group that had so successfully monopolized the price of beef in British Columbia.

Thomson had previously discussed with Ward the advisability of dissolving the beef syndicate. On 18 February 1884, Thomson and Greaves proposed to Ward that the two of them, plus perhaps Charlie Beak, purchase the remaining cattle owned by the syndicate at $35 per head. They would pay $50,000 in cash down, and the rest as the cattle reached market, or at the latest by 31 July 1885. Ward showed this plan to the three other members of the syndicate, and Pemberton responded unhesitatingly: "If all the subscribers are agreeable . . . write a line without waiting, accepting, as if delayed Thomson & Greaves may withdraw and not repeat so favourable an offer."

To calculate the purchase price, Pemberton jotted down a few rough figures. Allowing for 4 per cent losses, there were 3,662 cattle for sale at a value of $128,170. For the six syndicate members who two years before had each invested $5,000 in a "cattle speculation" this was indeed handsome profit—they had quadrupled their subscriptions. Even allowing for slightly lower values during the two operative years, the "corner in beef" had exceeded "a cool $150,000" profit by a handy margin.

All subscribers were agreeable to the February proposal and on 26 March 1884 the syndicate sold its stock to Thomson, Greaves and Beak. Out of the $50,000 they paid in cash, Thomson received a large remuneration for his services rendered as syndicate secretary-treasurer. On 1 April, each of the original six received his first instalment of the sale figure—$8,133.

Among them, Thomson, Greaves and Beak now owned over 8,500 head of stock, some scattered over the country at the ranches where Greaves had initially purchased them, the rest in the Nicola Valley grazing on the 8,000 acres of deeded land that Greaves and Beak owned around Douglas Lake. Thomson, Greaves and Beak considered that Ward's influencial position in the business world of the young province could prove beneficial in their plans. They soon made him see how attractive a joint stock company could be.

On 4 June 1884, these four men, Beak, Greaves, Thomson and Ward, formally agreed to operate a cattle ranch at Douglas Lake. As Beak and Greaves were already equal partners, each was to sell half his interest in the lands and stock of the partnership to one of the two Victoria men. As soon as Thomson and Ward had paid for their quarter interest, the four would incorporate "a joint Stock Company... for the purpose of carrying on a farming, stock raising and butchering business."

Douglas Lake Ranch began business that June with Charlie Beak as interim manager. The ranch accounting books that began with the March 1884 purchase of syndicate cattle show the continuation of the syndicate's buying practices. Every other month, Douglas Lake Ranch purchased hundreds of head of cattle, leaving most of them on the seller's ranch until needed.

Greaves and Beak now strengthened Douglas Lake Ranch's land base. In addition to their 8,474 acres fanning out from John Douglas's original preemption, the two ranchers amassed more land: 500 acres here, a half section there. These purchases advanced Beak's idea of acquiring bottom land capable of raising sufficient winter feed. That summer, Beak's young men from Purton fenced in the new land. The thousands of head of stock with their motley collection of brands— ///, ₵B, 7/, HZ, ₩ —were already ranging on Crown land and there they stayed until the frosts and snows of the fall of 1884 drove them to the lower elevations of the ranch. The bunchgrass on this deeded land provided winter forage.

Leaving Greaves to manage affairs at Douglas Lake, Charlie Beak left in June 1885 for Victoria, where he opened a butchering outlet for the ranch beef. Beak's London Market at the corner of Yates and Douglas streets became the eighth butcher store in Victoria. Within the first six months of business it generated meat sales of $57,500.

Douglas Lake Ranch was supplying not only the London Market with beef, however. After 16 months of steady purchasing, the four partners had spent $170,000 on close to 8,500 cattle from other Interior ranches. In June 1885 they began reselling 5,000 head to several markets: Andrew Onderdonk, then nearing the end of his CPR contract; the Van Volkenburgh brothers' stores in New Westminster and Victoria; Robert Porter, John Parker, George Black, W.B. Townsend and Goodacre & Dooley, all butchers of Victoria. There was an immediate yield of $150,000. These heavy cattle purchases and beef sales continued for many months, and though Greaves tried new markets, his business relationship with Robert Porter strengthened over the years.

Early in 1885 the purchase of three parcels of land around Chapperon Lake—one from the Crown, one from Pike Richardson, and one from Laurent Guichon—more than doubled Greaves and Beak's holdings by an additional 14,000 acres. When the fall colours splashed upon the wild rose bushes, the red osiers and the pockets of aspen, the ranch cattle began to drift home and Greaves's men separated them into two herds in preparation for the snowy winter, one herd to stay around Douglas Lake and the other to go to Chapperon.

It was at this time that J.D. Pemberton wrote to Ward. "While thanking you exceedingly for offering me a[n] . . . interest in the concern I must say for me it has little attraction. If I had money unemployed I wouldn't risk a dollar in it much less borrow from the Bank to invest in a venture in the success of which I have so little confidence."

Frightfully cautious for himself and suspicious of others, Pemberton was not willing to bestow on Douglas Lake Ranch his planning ability, business energy or additional capital. He queried the sense of having management in "the hands of one or two uneducated persons," criticized the poor way in which the syndicate books had been kept, and hinted that the ineptitude was deliberate to cover up misdeeds concerning the income from beef sales and counts of cattle on hand. He doubted that the butchering enterprise could be successful because of its high operating expenses, and was certain that emergencies on the ranch would persistently require untold amounts of fresh capital.

Ward's lasting trust in Thomson gave him confidence in those whom Thomson had grown to trust: Greaves and Beak. A ranching partnership calls for an unusual amount of faith because of the physical difficulties of rounding up and counting every head of stock on the property and ensuring that they remain alive and available for market. The allegations that the skeptical Pemberton had voiced did not sway Ward from continuing to view his partners and their enterprise favourably.

Pemberton was undoubtedly also cautious because of the sudden absence from the market of the thousands of beef-eating railway construction workers. Just two weeks before, on 7 November, Donald A. Smith, one of the largest CPR shareholders, had driven home the last spike of the Canadian Pacific Railway at Craigellachie, British Columbia. Every rancher in the Interior would suffer from this collapsing market, just as they had between the end of the Cariboo gold rush and the beginning of railroad construction. Even so, it was strange that Pemberton should have regarded the new venture so adversely, for that very year he had built himself a brand new mansion of 20 rooms on his Gonzales estate at a cost of $10,000, half of his share of the disbursement figure from selling the cattle syndicate that Greaves had managed.

Greaves jotted a note in early December to enlighten Thomson on ranch progress. "We are very busy now moving cattle on to the winter run—I will come down as soon as we get the cattle in the field—there is 160 cattle on the way down—we will start 150 head tomorrow."

Ward had asked to know "the number of cattle on the Douglas Lake Ranch" and Greaves figured there were "about 9,000 head—we will brand 2,500 calves this year." He was deferring branding until the cattle were on the winter grounds. Greaves further reported, "There is 30 miles of fence, buildings & corrals, sheds, everything in good shape to carry that number of cattle." Their band of horses, 60 in 1883, now numbered 100. Fences surrounded all the land, "excepting about five hundred acres—will fence that next spring." Hay had been cut too, and Greaves added, "We have feed enough to carry us through average winter with safety." As to the prices for next spring's marketable cattle, the cattleman guessed "23 dollars all round...all our cattle number one."

With Greaves as his advisor, Thomson drew up an evaluation of Douglas Lake Ranch at cost as of 1 May 1886. He valued the 9,000 mature cattle, 1,500 yearling heifers and 1,500 yearling bulls at $310,500; the 100 horses at $100 each; land at $39,000; surveying expenses at $1,500; and fencing at $7,000. With sundries, he placed the total ranch value at $378,000, and it was a quarter of this figure that Ward and Thomson finished paying that month in order to receive their quarter interest in the ranch.

Two years after agreeing to form a joint stock company, Greaves, Beak, Thomson and Ward incorporated "The Douglas Lake Cattle Company, Limited Liability" on 30 June 1886. Its objects were "to acquire lands in British Columbia for the purpose of raising cattle and horses, to buy and sell lands, to buy and sell horses and cattle, and to carry on the business of farming, stock-raising, butchering, and all matters incidental to the above purposes or any of them." The capital stock of the company was $400,000, divided into 400 equal shares. There were three trustees: Greaves, Thomson and Beak. Victoria was "the principal place of business of the Company," and the legal documents were drawn up by M.W. Tyrwhitt Drake.

Just two days before, the first trans-Canada train had left Montreal to arrive at Port Moody a week later on 4 July. The cattle company that had come into being because of the railroad's construction incorporated during the week of the first transcontinental run.

The final indenture that Beak, Greaves, Thomson and Ward drew up on 8 September 1886 gave them an equal interest of 100 shares each in Douglas Lake Cattle Company.

That same day the company took over title to all the Douglas Lake lands, 22,765 acres of them, plus the cattle, horses, equipment and improvements.

The 1882 cattle syndicate had planted the seed of The Douglas Lake Cattle Company, Limited Liability; the Beak and Greaves partnership had supplied the nourishment for its growth; the Thomson and Greaves proposal in February 1884 had brought it into being; the June 1884 agreement gave it strength; the June 1886 articles of association and the September 1886 indenture had bestowed final blessings. Douglas Lake Cattle Company was ready to face the future with four owners: Beak, Greaves, Thomson and Ward.

CHAPTER SEVEN

*We learn that W. Munro, of Douglas Lake, has sold his ranch to J.B.
Greaves, the great cattle buyer, who will in a few years own all the
ranches in the Douglas Lake country. Well, they might fall into worse
hands.*

Daily Colonist,
8 August 1886

When The Douglas Lake Cattle Company, Limited Liability,
incorporated in July 1886, it was already a large outfit
running 12,000 head of Durham and Aberdeen Angus cattle
on 35 square miles of deeded bunchgrass land around
Douglas and Chapperon lakes. In its 24 months of operation
it had already grossed $447,000 from selling syndicate
stock, Douglas Lake Ranch stock, and cattle purchased
from other ranches. And the other three owners made it
clear that Joseph B. Greaves, as its first fully fledged business manager, was to make Douglas Lake an even larger
and more viable ranch by plowing back profits.

If a small stockraiser in the drybelt Interior found it hard
making his place pay, he could always sell out to his neighbours and turn his hand to something else, as Pike Richardson had done when he sold his land to Beak and Greaves, and
opened a stopping house. Men with unlimited spirit yet
limited capital moved into the land, preempting, proving up,
selling out and moving on; and their land would enlarge their
neighbours' holdings.

Douglas Lake Cattle Company's goal was to hold on to its
land and to buy up all available acres. The land was there,
good bottom grazing land and benchlands stretching north
from Chapperon Lake, west and south from Douglas Lake,
and some unclaimed between. Beyond the bunchgrass in
every direction rose high ridges of timber with a grassy forest
floor, thousands of acres of excellent summer grazing. The

vast valley of the Upper Nicola River contained all the features necessary for a great ranching enterprise. J. B. Greaves had the grit and tenacity necessary to create such an enterprise.

That first year of Douglas Lake's operation tested Greaves and his fellow shareholders. It began surreptitiously during the haying season when heavy rains fell on the windrowed hay, leaching out its quality and delaying its harvesting. Greaves was meanwhile too busy to worry about the hay—putting up winter feed was still considered almost a luxury. He was establishing the Interior cattle market by paying $35 a head for three-year-old steers and $30 a head for two-year-old steers and fat cows, buying more land, or shipping beef down to Beak's butcher store. But the winter was impossible to ignore. From the middle of January, blizzards raged throughout the Interior, dropping blankets of snow to cover even the tallest bunchgrass. For over a month, storm followed storm and haystacks shrank until nothing remained to feed the once grass-fat cattle.

The spring of 1887 revealed thousands of Douglas Lake carcasses. Yet with spring came renewed vigour to the land, to the surviving cattle and to the men who were caring for both. Greaves offered "six bits apiece" for every carcass hide that the Spahomin Indian women skinned. The company began rebuilding its herd. In those early years, the number of cattle bearing ||||, an expansion of Beak's old brand signifying four owners, reached an all-time high of 18,000 head.

The riders, farmers, fencers, teamsters and chore boys caring for the cattle at Douglas Lake were predominantly Indians from Spahomin: Okanagans, Athapaskans and Thompsons who worked as they or Greaves needed. The shrewd Yorkshireman treated his workers paternally, understanding their emotional and physical strengths and their indifferent work habits. The Indians were natural riders, strong fencers and fine teamsters but casual farmers. However, the Chinese railway construction men whom Onderdonk had laid off in 1885 and who had not returned home were fastidious irrigators, brush cutters, gardeners, and steady cooks. Eager for more money to send back to their families, these serious, blackpigtailed men hired on with many Interior ranchers, including Greaves, who respected their wish to remain distinct from the rest of the crew.

From the start, Greaves sought good men to hold the responsible jobs: Richard Murphy, herdsman; J. Theodoro and William Lunn, farmhands; D. McDonald, carpenter; Tom Jones from North Wales, who was to rise to the status of foreman during his 24 years with the company.

There was Jack Whiteford, nephew of Mrs. William Palmer, who in his 20-odd years at Douglas Lake proved himself an excellent cowboy, often taking charge of the beef drives to Kamloops and occasionally acting as cowboss.

Frederick King, one of the three boys Beak had brought over from Purton, stayed on under Greaves after reaching the end of his four-year contract, and became renowned as a top horsebreaker during his 20-year stint. At Grande Prairie, he was injured by a runaway team and though the company paid all expenses for him to go to the Mayo Clinic, King died a young man in 1918, after five years in a wheelchair.

There was Merino, who ran the Howse Meadow and the Morton as a wintering unit. A.E. Howse, proprietor of general stores at Nicola, Lower Nicola and Granite Creek, had sold his land south of the Douglas Lake wagon road to the company to repay a mortgage. The Morton was the superb winter and spring range that Charlie Beak had lost to Alfred Morton over a technicality. Morton refused to sell to Douglas Lake and it was not until 1890 that Joseph Guichon purchased the title on Douglas Lake's behalf. Having to name such good land after a "miserable claim jumper" always bothered Greaves.

Merino had come up from Mexico during the gold rush as an expert packer. Now he lived in a small cabin adjacent to the Morton, cutting the wild hay of Howse Meadow in summer and rationing it out during the short winter. The winters were short on this part of the ranch, for the Morton was one of the first ranges in the valley to become bare of snow each spring, and cattle could graze there as early as March. The Morton was the kind of hay-saver for which every rancher yearned.

But the most important ranch hand then was Joseph Payne, an American. Payne was experienced in all aspects of cowboying, from tending sick cattle to driving cows and calves. Cowboss for nearly a decade, he maintained the good cattle husbandry that Beak and Greaves had practised. He was a good teacher, and the Indians who worked with him adopted his methods of getting cattle out of the bush, driving, holding and sorting cattle for market and for weaning, branding cattle, and giving special care to the sick ones. Payne married while at the ranch, but he never brought his wife to live with him; the cow camps were too rough for a white woman. Instead, he visited her frequently when he drove stock to the railhead at Kamloops.

Once more Joseph B. Greaves bought land, acquiring parcels from the Crown and from settlers to fill in the spaces between the parcels already owned. He added, in 1887, another 7,500 acres for which the four shareholders had to borrow money from the bank at heavy interest rates. This rapid absorption of so much of the Upper Nicola Valley alarmed the ranchers of the Nicola River watershed area, for each nursed his own dreams of expansion and it seemed that the big company was going to acquire all the land. In May 1887, 58 landowners petitioned the chief commissioner of lands and works in Victoria to set aside a 25-square-mile block of land south of Nicola Lake as a commonage. On 18 May the Lumbum Commonage came into existence, named after A. W. Lundbom, a settler who had lived there in the 1870s. John Clapperton, the government's representative in Nicola, acted as assessor and revenue collector for this commonage and it was soon filled with local cattle.

Greaves's land purchases the next year were only 1,000 acres, most of which enlarged the northern boundaries so that the ranch now encompassed Fish Lake, north of Chapperon, the small body of water known to the Okanagan Indians as Head Water Lake.

At the end of that year, another commonage comprising almost 25 sections surrounding John Hamilton's old "Mountain Ranch" emerged between the Lumbum and Douglas Lake. The neighbouring ranches, including Douglas

Lake, that used it from the start called it the Hamilton. Still uneasy about the growth of Douglas Lake and the potential lack of grazing land, the Nicola Valley settlers petitioned for a third commonage. Known as the Marsh Meadow, it comprised 29 square miles north of an imaginary line drawn between Douglas and Chapperon lakes—an area in which Greaves had already contemplated purchasing land. It was an obvious blocking ploy, but as Douglas Lake's stock made up the largest proportion of cattle in the valley, they made good use of the Marsh Meadow Commonage from the beginning.

For some time, Greaves had heard rumours of a British breed of beef cattle that could withstand the rigours of such cold winters as that of 1886-87, when so many of his Aberdeen Angus and Durhams had died. Already Eastern Canadian and American stockmen had imported this breed with good results. In 1888, having bought less land than in previous years, Greaves brought in some Hereford cattle, the first to the Nicola Valley. These red-bodied, white-faced, straight-backed animals came from the Quebec herd that Senator M.H. Cochrane had founded in 1880.

Over the years the Hereford breed proved itself by calving easily, growing fat on good grass, and faring well in winter. It replaced the Angus cattle on the ranch, and equalled the Durhams, or Shorthorns as they were called. The purebred Hereford had excellent front quarters and great hardiness, but was somewhat lacking in its hindquarters, whereas the Shorthorn had good hindquarters. Crossing of the Hereford and Shorthorn breeds was essential, and improved the hardiness of the Douglas Lake strain.

Just as those first Herefords were arriving at Douglas Lake, Charles M. Beak returned from an extended trip to Great Britain, having previously closed down the London Market. During his trip he had attended the Glasgow Spring Stallion Show and had purchased the winning two-year-old, The Boss, plus three other Clydesdale stallions of the best breeding. He had had to pay $4,700 for The Boss, his sire being the most successful breeding horse in Scotland.

Greaves fairly exploded. Forty-seven hundred dollars for a stallion that he could only breed to rough saddle and pack horses seemed sheer lunacy. To support the land purchasing campaign and the Hereford purchases, Greaves was having to scrape and save as much as possible. He argued that money should first be spent on haylands before the partners acquired such fancy horseflesh. So conscious of economies had he become that in the Ranch House, chains held the tin mugs to the tables to avoid petty theft. Beak himself had occasionally inspected the drained mugs for undissolved sugar to ensure that no one was wasting provisions. If he found any sugar, the men responsible were the first to be let go at the end of the haying season.

But Beak was certain of the soundness of his purchases, and slowly brought Greaves around to his way of thinking: so much so that the next year Greaves ordered six registered Clydesdale mares from Sorbey of Guelph, Ontario. Once the Clydesdale stallions' first get were on the ground, Greaves began to understand Beak's enthusiasm. Soon the stallion colts were spoken for even before they were born, and Greaves was admitting that the purchase of The Boss was one of the best investments the ranch had ever made. These heavy draft horses provided all the locomotion for the farming at Douglas Lake and their name spread. Eventually, the value of the Clydesdale sales became so large that it almost covered the ranch expenses for a time, leaving the sale of cattle as pure profit.

Greaves, already a good judge of horses, became a fine judge of these draft animals and a great hand at raising them. He never allowed his men to cut a stud under the age of two, and allowed them to halter-break a colt only using hip ropes, a long heavy rope brought around the horse's rump, across its back, over the shoulders, through the halter and then to the barn wall. With the hip ropes short and the halter long, when the colt strained away from the strange constriction around its head, it could not kink its neck and become useless for pulling; the hip ropes would take the strain. Each evening Greaves would visit the yearling colts in the barns, telling them to "lie down and grow fat for me."

The eight lads who brought the stallions over from Scotland stayed on as farmhands as their foreman, John

Blackwell Baldwin, worked with the Clydes. Later he moved south to Minnie Lake, taking up land there and marrying Lavina Earnshaw, daughter of the pioneer Minnie Lake settler Byron Earnshaw.

A better forge was needed for the Clydesdales and was erected in the centre of the Home Ranch yard. Each day the distinctive smell of heated iron and seared hooves assailed everyone passing through the yard to the barns and corrals. From time to time the air rang with the rhythmic metallic strikes of Bishop, one of the earliest blacksmiths, hammering out a white-hot horseshoe or gate hinge. As well as shoeing horses, he sharpened plowshares, overhauled mowers, and rimmed many wooden implements with steel: wagon wheels, single trees, double trees, spreaders, and neck yokes. Every piece of horse-drawn equipment from the binders to the grader received attention from Bishop's hand. The black-smith and his shop rose in importance as the ranch took better advantage of the capabilities of its draft horses.

In 1889, Greaves made the first purchase of Minnie Lake land when he bought from the Crown almost 2,000 acres, which became known as the Wasley because it neighboured land that Samuel Wasley had preempted. Douglas Lake Cattle Company land now stretched 27 miles from northeast to southwest, though there was still much land between that was either vacant or occupied by others. The future outline of Douglas Lake was beginning to take shape.

Another 2,000 acres added to the company holdings that year, some coming from Jack Douglas and his uncle. Not long after Jack sold the company his ranch that lay between Douglas Lake and the Marsh Meadow Commonage, he heard sad news of his uncle. After selling his first Douglas Lake land to Beak in 1883, Old Douglas had gone mining at Rock Creek in Boundary country. Each year, he returned to Cloverdale, Sonoma County, California, "under the treatment of Mrs. Preston, who claims the miraculous power of curing the sick by the laying on of hands." The good lady was unsuccessful in curing Old Douglas's consumptive condition, however, and on 18 March 1889 he died there.

Jennie Douglas, his estranged wife, read the newspaper report of Old Douglas's death and promptly hired lawyers to claim the $5,000 of certified cheques in Old Douglas's belongings that were to go to Scotland, where his brothers and sisters had his will. Once in California, Mrs. Douglas, "one of Vancouver's most estimable Ladies... who has long kept a fashionable boarding-house on Homer Street," established her identity. She received the $5,000. What she did not know was that some time after 1883, Old Douglas had purchased another 500 acres of land at Douglas Lake. The title to this land reverted to nephew Jack, who at once sold it to Douglas Lake Cattle Company.

Jack then left the lake that had taken his and his uncle's name, and put the $7,500 that Douglas Lake had paid for his land and 300-head herd into 3,360 acres of land situated halfway between the young town of Revelstoke and the Rockies. He died in Vancouver in the 1920s. Jennie Douglas operated a boardinghouse in Portland, Oregon, and Douglas Lake never heard of her again.

With the end of the year 1889, the shareholders of Douglas Lake Cattle Company reached the end of their arduous financial struggle. Just as Pemberton had suspected, they had been underfinanced from the start, and had borrowed large sums of money from the bank—paying $110,000 in interest alone. They had absorbed another $90,000 of bad debts in that same period. Now Douglas Lake Cattle Company could pay its own way, the horse and cattle sales equalling the ranch operating and capital expenses and at times exceeding them handily.

The only event in 1889 that might have alarmed J. B. Greaves was the arrival of grasshoppers. Instead, the ranch benefitted. The *Inland Sentinel* reported: "A gentleman from Nicola states that the grasshoppers are dying off, and that the devastation caused by them is much exaggerated. The ranchers are much discouraged, however, and are constantly selling their farms to Douglas Lake Cattle Co. It is only a matter of time, our informant states, until Nicola Valley is entirely in the possession of this wealthy company. The latter, however, pay the farmers satisfactory prices for their holdings."

This was not the first time that grasshoppers, close relations of the biblical crop-eating locusts, had swarmed the Nicola Valley in such numbers, for the Spahomin Indian, Old Tom, remembered such outbreaks during his youth. But the 1889 infestation was the first that had affected ranchers, and it was an omen for what was ahead.

It came to the attention of the B.C. Lands and Works Department in the summer of 1890 that the Marsh Meadow Commonage had not been much used. In November of that year, the government threw open the 18,553-acre commonage, and despite its apparent unpopularity, the rush was on. By enrolling 19 individuals to sign as preemptors, and so avoiding the technicality of being limited to preempting only one title at a time, J.B. Greaves copied a technique which many cattle kings and land companies had used. The Lands and Works Department, eager to dispose of the land and to encourage its improvement, turned a blind eye to this abuse. J.B. Greaves succeeded in acquiring just over 10,000 acres of the old commonage while other settlers in the valley took the remaining 8,000 acres.

Only when Joseph Lauder picked the Marsh Meadow land to preempt did J.B. Greaves realize he had blundered: a corner of the new Lauder land was clearly visible from the Douglas Lake Home Ranch yards—to Greaves's eye, a blot on the magnificent horizon. How many times afterward he wished that he had claimed that piece for Douglas Lake! Fortunately he never allowed his mistake to mar the neighbourly relations he enjoyed with the Lauders.

Another preemptor of the cancelled commonage land was to become a thorn in the flesh of both Greaves and Douglas Lake as time passed. Alfred R. Goodwin, an immigrant from Aylsham, Norfolkshire, was "generally popular, hot tempered, a hard worker and very ambitious," and it was this very ambition which finally caused problems.

Two young men who were to become long-time employees at Douglas Lake joined the ranch labour force at this time. Ed Godley sold his saddlery shop in Calgary before arriving in

1890. Except for a brief spell with the Cariboo Express, he stayed in the Nicola Valley for the next 55 years, at first moving around the different ranches to attend to all the harness and saddle making, later staying closer to his headquarters at Douglas Lake. The other young man was 23-year-old Joe Coutlee, son of Alex Coutlie, whose Nicola Valley ranch Ward, Thomson and O'Reilly had mortgaged some years before. Young Coutlee's career had been obvious from the age of ten when he had first worked cattle with Joseph Castillion, the Mexican packer who had once milked cows for Charlie Beak.

Late in 1890, Joseph B. Greaves estimated that of the 80,000 head of cattle throughout the province, the 10,000 inhabitants annually consumed 24,000. British Columbia was just self-sufficient in beef production. Of the 80,000 head, 30,000 were grazing in Nicola Valley ranches, over a third of that figure on Douglas Lake Cattle Company land. The joint stock venture embarked upon by Greaves, Beak, Thomson and Ward a few years earlier had risen to a company of prime importance in the province. By 1892, it had doubled in size, from 22,765 deeded acres at incorporation to over 47,000. The shareholders could boast about their great Herefords and Shorthorns, their money-making Clydesdales, their irrigated haylands, and their loyal employees. But the whole venture dispirited one of the four owners.

Charles Beak and J.B. Greaves had been unable to agree on the future of Douglas Lake for some time. Beak was of the opinion that the ranch should purchase only bottom land, while Greaves was certain that the only way a ranch could become great and remain viable was by having a solid foundation in the land, both bottom land, sidehills and timbered ranges. They were demonstrating that given the same land, cattle, horses, manpower, and market potential, no two ranchers would ever react in the same way, for like most entrepreneurs, ranchers are distinctly individualistic.

Having moved to Salem, Oregon, Beak sent a certificate of sale, dated 18 January 1892, for his 100 shares in the company to the three remaining owners of Douglas Lake Cattle Company. One hundred and ten thousand dollars was Beak's sale price: half down, and the rest spread over the next three years at 3 per cent. Greaves, Thomson and Ward took up the offer.

Beak was to spend the remainder of his life quietly. In 1895, he sold his 2,000-acre farm in the Willamette Valley, Oregon, and returned with his family to Purton, Wiltshire. There he purchased Purton Manor House, an impressive stone mansion that stood between one of England's largest tithe barns and the twelfth-century Purton Parish Church. In June 1900, Purton parishioners laid him to rest in the churchyard alongside his parents, brothers and sister.

Though Greaves and Beak had some differences, the Douglas Lake manager had relied heavily on his partner's good judgement. Together they had realized a dream in building the ranch.

Greaves was already 61. His trim moustache and goatee were fading to a snowy white, and the companion of old age—rheumatism—was afflicting his joints. Yet over the next 18 years he would continue running and developing the ranch he had helped to found with even greater efficiency.

CHAPTER EIGHT

To get a horse with | | | on the left shoulder is just the same as buying
a sack of Brackman & Ker's rolled oats, the quality is always there.

Anonymous, "Douglas Lake Cattle Co.,"
1899, G.D. Brown Papers

The three-way division of Douglas Lake Cattle Company
shares between Greaves, Thomson and Ward in January 1892
all but coincided with the end of financial hardships for the
ranch. Under Greaves's management the acreage increased
and the ranch operation intensified.

Again, a once-in-a-decade bad winter occurred. It began
fairly normally in 1893 with a heavy snowfall and then a thawing Chinook wind. Before the Chinook had melted the snow,
temperatures dropped and turned the snow to glare ice that
flattened even the sturdiest bunchgrass and spelled treachery to
anything on foot.

Hundreds of Douglas Lake cattle slid down the hills, breaking legs or necks, or both. Before the winter was through,
some of the surviving cattle were so weak they had to be supported in hoists. For many, slaughter was the only answer,
and by late spring 8,000 head had died. Using a derrick, the
crew piled the carcasses and set fire to them. In 1893 over 2,500
calves had been branded; in 1894 the tally was 300. If the
ranch was to withstand such fierce winters, the natural hay
meadows would have to be improved to grow as much hay as
possible.

One of the first wild hay meadows to receive attention was
the Big Meadow of 200 flat acres south of Chapperon Lake.
Greaves dammed Chapperon Creek and placed a sluice gate at
the outlet so he could flood the meadow 3 feet deep to irrigate
its native grasses in spring. Within a few years, 800 acres at the
Home Ranch, Chapperon Lake and Minnie Lake were under

cultivation and yielding 150 tons of grain and 1,600 tons of hay annually.

The ranch acquired some new land after John Gilmore burned to death along with three of his five children in their two-room log house close to the town of Nicola in January 1894. Before his death, John had borrowed $2,200 from relations, who now sued his widow, Ellen, for repayment. She did not have the necessary cash, so she auctioned off all the ranch's cattle, horses and implements. The land—2,500 acres of hilly land and shore around the west end of Nicola Lake— was all in Greaves's control by 1895, but lying as it did many miles west of the nearest Douglas Lake land, it could not be used immediately for growing hay or grazing cattle. James Gillie, William Pooley, and Ellen Gilmore rented the land from Douglas Lake in three separate parcels.

The Gilmore land, so close to Nicola, could become valuable, for the CPR had promised in early 1894 to build a branch line from Spences Bridge into the heart of the valley. The town of Nicola could hardly wait to be "the terminal city of the Nicola Valley Railway," and many were planning for that time.

Although the range wars in the United States had no counterpart in the Nicola Valley, the Interior ranchers were definite in their stance against sheep, and new legislation—the Range Act of 1893—declared the valley a sheep-free area.

Horses, too, had their critics so that the *Inland Sentinel* could report in August 1895, "There are now upon almost every range of the Interior very many more horses . . . than are likely to be required for work or the saddle for many years to come." Though the Hudson's Bay Company at Kamloops had sold off a large remuda of brigade trail horses in 1878, these horses' numerous offspring, well mixed with Indian cayuse, roamed the bunchgrass sidehills at the north end of the Nicola Valley. The Kamloops newspaper continued with some advice: "The horse is a noble animal which one does not like to see slaughtered, but in the cold utilitarianism of the day his carcass is more valuable as canned meat for European soldiers than as an embellishment of the hill side."

The 700 heavy horses that were "embellishing" the Douglas

Lake ranges at Minnie Lake at this time were in fact beginning to pay their way admirably. In order to stabilize the size of his expanding remuda, Greaves started selling as many as 100 horses annually, some as weanlings just off their mothers for $50 each and some as two-horse teams for $300. As the good reputation of the Douglas Lake Clydesdales spread, their demand and in turn their asking price increased.

The $8,000 or so from these horse sales every year provided half the profit that the partners annually divided between themselves. The rest derived from the selling of over 4,000 cattle—three quarters being ranch-raised steers and cows, one quarter being Okanagan cattle, bought and fattened. About $8,000 profit remained from these sales after ranch running expenses of $27,600, bad debts and bank interest of $12,000, and land and cattle purchases of $58,400. This period proved to be the most financially successful in the life of the ranch and its profitability was due in part to the descendants of The Boss.

Douglas Lake's many land transactions, and its increased payroll with larger haying and cowboying crews, had required more accurate bookkeeping for some time. Also the small store, just opened at the Home Ranch to sell such goods as tobacco and men's overalls, needed a storekeeper. A temporary clerk was hired in 1894: George Bent, a semi-retired miner, packer, storekeeper and Minnie Lake rancher who had cowboyed with Beak and Greaves. In 1895 43-year-old Robert Beairsto, a Prince Edward Islander, seventh-generation Maritimer, and qualified teacher, took over this job and also began caring for the studs and the chickens.

Bob took his work very seriously; if it happened that he had customers in the store when it was time to collect the eggs or water the studs, he would say, "Do you want to wait inside or outside?" Off he would go, snapping a padlock on the door as he went, his patient customers locked inside or out.

The small store began by selling immediate requirements—toothpaste, camphor ice, tobacco, candy, oranges and playing cards. As Greaves paid Indian labour in kind, the inventory included clothes—men's "Denham" overalls, gloves, cowboy hats, flannel and wool drawers and shirts, mufflers, handkerchiefs, travelling rugs, boots; a man could bundle up for winter for just $16. And for the Spahomin womenfolk, who worked as hard as their men did on the ranch digging potatoes, pitching hay and cooking, there were velvet shawls, ladies' coats, underskirts, knitted hoods, yarn, and yardages of dress goods. Black corded silk ribbon was available for Ah Kwong, Ah Loone, Ah Lee, Ah Gee, Ah Bob and the other Chinese on the crew to make into ties.

For the farm work there were scrapers, walking plows, spade harrows, rakes, mowers, reapers, binders, pumps, machinery parts, horse collars, wheelbarrows, picks, shovels, manila rope, hayforks, scythes, grain sacks and barbed wire and staples for adding to the 250 miles of fence. The 8-ounce duck that was sold for tents, and the ticking meant for mattresses, enabled the crews at far-flung hay meadows to make themselves outdoor homes.

Beairsto brought in leather, snaps, toggles and buckles so that Ed Godley could make draft harness and cowboys' "rigging" (saddlery). Some ready-made saddles, spurs, and chaps came in from Kamloops and Calgary.

Anxious to save by buying wholesale, Greaves instructed Beairsto to enlarge the inventory at the store to include such groceries for the Ranch House and the various camps as cured bacon and hams; rice; dried apples, prunes, peaches, beans, and codfish; and root vegetables and green beans grown from seed by the Chinamen.

Other staples included brooms, lamp wicks, candles, matches, lanterns, blue-mottled soap, flypaper, coarse salt, preserving jars, coffee mills, kettles, washtubs, milk pans, dish washing pans, tin cups, tin plates, white cups and saucers, steamers and washboards.

One or two Indian teamsters were sent to Kamloops each week during spring and fall to collect from the Hudson's Bay Company groceries and any freight that had come in, the round trip taking five or six days. The teamsters occasionally hauled both ways when Douglas Lake had hides or vegetables to sell. Sales from the store annually grossed around $7,000.

The ranch's general store, which on a busy day might have over 30 customers, including patrons from neighbouring land, became a centre of activity, with horses and wagons tied to the hitching post outside. From the Indian crew, Caprian and Dry Wheat's Boy might report on the seven miles of fence and "krall" (corral) they had built and be credited at the rate of $30 per mile. Ed Godley might exchange 12 quirts he had made for $12 cash. Grub orders for Chapperon House, Morton Place and Minnie Lake Camp would come in. A Spahomin woman—a "kloochman," Beairsto would call her—would trade a bundle of new buckskin moccasins or a ling cod caught from Douglas Lake for groceries. The store opened early and closed late, weekends as well as weekdays.

In 1896, after 32 years away, William Curtis Ward and his wife returned to England to live in Harbourne House in Kent. Ward, who had become his bank's superintendent in London, formally adopted his *ex officio* seat at its Court of Directors late in August 1897, replacing Sir Charles Tupper, whom Canadians had elected prime minister.

C.W.R. Thomson too was coping with changes that year, for in April the B.C. Electric Railway Company Limited had become incorporated, forcing the Gas Company to compete with electric lighting by lowering its prices and increasing its efficiency.

In May, J.B. Greaves's life was in danger, according to threatening notices posted near the ranch which stated that "Felix or 'Dry Wheat Joe' proposed shooting him on the earliest opportunity." Complaints led to a preliminary hearing at which several Indians swore that Felix had tried to persuade them to kill J. B. Greaves and steal a band of horses, which they would run across the American line. Justice of the Peace Clapperton committed Felix for trial at the June assizes for "inciting to kill."

If Greaves were on the warpath, hat tilted almost to the bridge of his nose, his men would call him "Old Danger." But normally they referred to him as "The General" or "The Old Man." Said a visitor to the ranch:

Mr. Greaves is the manager, and a capital manager he is. But of all the conservative men in the world he is one of them. No man on earth knows what he looks like unless he has seen him, for he would

sooner fight the Boers than have his picture taken. He has successfully evaded every kodaker who has ever visited Douglas Lake. He
was nearly caught at one time—a snap shot revealed the back part
of his head and handkerchief that he usually wears as a collar. He is
one of the most unassuming men in the country, yet all who know
him like him. And what he does not know about running a big ranch
would be hard to figure out.

To Greaves, running a big ranch meant taking equally good
care of the men, horses, cattle and land. He was as one with
his crew. A Thompson River rancher with cattle to sell invited
Greaves and his men to stay for lunch. When the men had
stabled their horses, they drifted into the house and sat down
in the parlour. Finding them there took the rancher very much
aback. "I never allow my crew in the parlour, Mr. Greaves,"
he said. "All right, boys. Out we go," said Greaves, and he
followed them out.

Each day, the ranch followed a strict routine. Everyone had
to be up by 4:30 A.M., giving them ample time to "jingle"
(round up) the saddle horses and to feed, water and harness
the work horses. Greaves attended to his own team or horse
personally. At 6:00 A.M. everyone would gather for breakfast
in the Ranch House, the central building at the Home Ranch.
Besides the large dining room, this structure held the kitchen,
lorded over by a succession of Chinese cooks, and Greaves's
own bedroom and sitting room. By 6:30 A.M. everyone would
be at work.

At the evening meal, once all had seated themselves along
the table, the Chinese cook would place the roasts in front of
Greaves. He insisted on carving whether there were 10 or 40
having dinner, and, thanks to his butchering days, did so
quickly and efficiently, piling each plate high with steaming
beef. If a crew member complained that his helping was too
large, Greaves would tell him to eat up, for without eating, a
man could not work, and a man who could not work was no
good on the ranch. The crew member suddenly found a large
appetite.

With his good sense of humour and his love of storytelling,
the short evenings passed pleasantly in Greaves's company.
His was usually the last tale of the evening, and at 8:30 P.M.
prompt he would depart for bed.

A generous man with his liquor, Greaves willingly dispensed encouragement from his cases of Dewar's Scotch or from the two-gallon jugs that he ordered filled with aged Hudson's Bay rye at the Hudson's Bay Company store in Kamloops. At the end of any day of strenuous labour, or simply when the crew had pleased Greaves, the jug would pass around several times. If the job had taken Greaves and his men from home, he would usher them all into the nearest bar and stand everyone a drink, Indians too, which was illegal.

On a cold day's ride, Greaves would warm himself with a draft from his hip flask. One day, in company with Johnny Bull, an Indian who was a heavy drinker, Greaves took his draft and passed it over. "Take a small drink, *hailo* strong." The small drink lasted Johnny an hour in the cold weather; then he reached down in his chaps, pulled out a bottle of Holland gin, and handed it to Greaves: "Here Grabes, take a drink, a big drink, *hailo* strong."

Around this time, a young Lillooet Indian, Billy Fountain, came to work at the ranch. During the Kamloops Fair, Greaves met the local policeman on the main street leading Billy to jail. Seeing that Billy was a little under the influence, Greaves asked the policeman why he was taking him in. When he heard it was because Billy was an Indian, Greaves replied with great aplomb, "Nonsense, he's a Pennsylvania Dutchman. Come along with me, William, and I will put you to bed."

Neighbouring ranchers were proud to call J.B. Greaves their friend; agricultural associations sought his opinion and appreciated also his generous financial contributions. Many of the smaller places greatly valued Greaves's financial help. There was no bank at Nicola, yet long before their steers and beef cows were ready to go to market, ranchers needed money to pay their taxes, wages and other expenses. It became customary for ranchers to ride to Douglas Lake, stay overnight, and leave in the morning with a cheque, having pledged their beef for fall delivery. Robert Charters turned these cheques into ready money at his store at Quilchena.

Greaves financed not only his neighbours but also many Okanagan ranchers who were badly cut off from a market. Every June he travelled there on horseback with about 20

riders, three horses each, and a chuck wagon. He would purchase up to 1,000 or more head of cattle and bring them to
Douglas Lake then or later in the year.

Seldom did anyone manage to "put it over" Greaves. One
time he sold a saddle and then forgot who had bought it.
When he sent his friends invitations to attend his annual
Christmas dinner, he enclosed a bill for the saddle with each
one. The owner of the saddle paid up.

Father Le Jeune was a French missionary Catholic priest
who devised a simple method of writing so that he could teach
12 different Indian tribes how to read and write and so learn
the gospel. A few Saturdays each year, he—or more com-
monly Reverend Mr. Murray—would travel from Nicola to
stay the night and hold a Sunday service. Following these services Greaves showed his gratitude with a handsome donation
of money or a roast of beef.

Indeed Greaves made everyone welcome, *if* they were kind
to their horses. Once a fellow showed up at the Home Ranch,
his played-out horse covered with lather. Douglas Lake's
manager informed him that if he ever came back he should
bring a tent, as he would have to camp at the far end of the
Big Meadow. His horse, however, would receive the best
treatment.

The big Klondike gold rush of 1898 attracted thousands of
miners to Dawson City, and to nearby Atlin. For the ranches
in the Nicola, including Douglas Lake, the new rush meant
increased horse sales. A Swede came by, looking for a good
horse for packing into the Klondike. His funds were low, but
Greaves grabbed the opportunity to advertise. He gave the
Swede a well-bred horse and even a two-wheeled cart, and instructed him to tell everyone he met along the trail to the
north that the horse had come from Douglas Lake.

Beef sales increased in proportion with the horse sales, and
in 1898 the ranch herd again grew with additions from the
Okanagan. One of W.C. Ward's sons, 23-year-old Francis B.
Ward, was on the drive that year, bringing back "250 mixed
cattle."

The Douglas Lake crew branded the Okanagan drive at
Chapperon Lake and as they finished at two o'clock in the

afternoon, a buggy containing two of the Sisters from the Providence Orphanage at New Westminster drove into the yard. They were on their annual trip collecting funds to look after the orphans in their care. Greaves ordered every man to put his name down for $1.00. He headed the list with $50.00.

It was in the late '90s that Joseph Payne switched from buying and shipping cattle for Douglas Lake Cattle Company to become a cattle buyer for Pat Burns. He later left Burns's employ to travel throughout the Americas, ranching in northern Patagonia, Argentina, and San Kula, Brazil. In 1914 he was back in Kamloops, running a billiard room. In 1924 he went to California. Though he spent only a short time at Douglas Lake, Payne made his mark there forever, in the superb cattle handling techniques he introduced.

A man who, if born in another setting, might have been an army general took Payne's place. Thirty-year-old Joe Coutlee became cowboss. Just as a general knows how long rations will last his men, so Coutlee could tell accurately how many head of cattle a field would hold and for how long. The rule of thumb that the Interior's drybelt country required between 30 and 40 acres to support one mother cow and offspring year-round held true at Douglas Lake, but it was of little use to Coutlee, who had no knowledge of acres as a measurement. He instead had his own method of "eyeballing" grass in a field, whether last year's bunchgrass on spring range, the delicate growth of the current spring's bunchgrass, the hard, yellowed range bunchgrass of summer and fall, or the bright green but water-soft timber grasses in the high country.

Except for two or three areas where he always ran camp horses or where sorting always took place, Coutlee guarded Douglas Lake's native bunchgrass well, realizing that if cattle or horses grazed bunchgrass plants during their spurt of first spring growth, they would not develop heavy enough roots to survive. Hence he separated the herd, as Payne, Greaves and Beak had done in the past, and kept each group moving to higher elevations through the spring so that the cattle grazed the previous spring's growth. By summer, all but the marketable cattle would be in the timber, thereby saving the range grasses for later use once November snows started moving the cattle down towards the ranch once more. These good range management practices have been followed to this day.

Coutlee knew his stock well: out of a herd of 13,000 Shorthorns and Herefords he could identify one that a cowboy had pulled from a bad mud hole two years ago, or one that was just a calf when it lost its tail to a coyote. When rounding up, Coutlee would rest his horse on a high knoll, surveying the cowboys to ensure that he had planned his manoeuvres well and that the boys were carrying them out just as faithfully. And always, after a long day, it would be Coutlee who would ride the biggest circle, often coming into camp last.

If Coutlee caught any of his men reading in off time, he would say that if there were time to read, there must be much more time for splitting wood for the cook. When his men wanted to go to town for a change of atmosphere, he would give them unbroken horses to ride. That way they were working—horse breaking—even on their way to a night on the town.

Coutlee, for many, many years to come, was the right man in the right place. From his French father, Alex Coutlie, came his charm; with his mother's Indian blood he could understand the Indians and get better work from them than any "white man" could. He preferred his crew to have as many Indians as whites, so that neither group would have its own way. During slack times Coutlee and his crew might all be drunk and incapable of work, yet when there was work to do, they did it quickly and efficiently. Coutlee was rehired as often as he was fired.

In April 1900, after 15 years' ownership, Owen S. Batchelor sold his 1,200-acre ranch north of Fish Lake to Douglas Lake Cattle Company. This land became the most northerly part of Douglas Lake. Batchelor's wife, Julia Bradley, had been Grande Prairie's first teacher, among whose first pupils had been J. B. Greaves's sons Joseph and Peter. The Batchelors remained in the Kamloops area the rest of their lives, Owen spending some time on the land but mostly following the love of his life, mining.

In England, W. C. Ward had been deeply involved in bank affairs, and in July 1900 he wrote to Peter O'Reilly to advise him of the amalgamation of the Bank of British Columbia with the Canadian Bank of Commerce: "I think the united Bank will make an important concern. . . . I fully expect that the shares of the Canadian Bank of Commerce will considerably appreciate in value within the next twelve months."

Always a man of modesty, Ward failed to point out that since 1884, the Bank of British Columbia had increased its capital to £600,000, while its reserve was at £100,000; that it had eight branches within British Columbia—Victoria, Vancouver, New Westminster, Nanaimo, Kamloops, Nelson, Sandon and Rossland—and that this progress had occurred under his managership.

Returning to British Columbia in December 1900, Ward found Thomson and Greaves selling fattened cattle to a new—to them—cattle buyer, Pat Burns. This Canadian of Irish stock owned packing plants and butcher stores in Alberta and British Columbia that did $1.5 million worth of business annually. Thomson and Greaves were also supplying their old buyer, Porter, to keep competition keen.

Cattle for market left the ranch from the Beak Place, as Greaves called the Morton, in drives of 160 to 240 head twice a month during summer and fall as their grass-finished condition dictated. The 40-mile trail ran northward cross-country through the Moore Field, the first overnight stop for the cattle and shipping crew being at Stump Lake Ranch. After grazing on the native grasses, the cattle bedded down for the night, herded by a few riders. The rest of the crew opened their bedrolls in any barn or outbuilding that afforded shelter. Only when the shipping drives dragged on into winter did Greaves buy hay along the route to feed the cattle.

Trailing north along the Brigade Lake trail, the drive spent the second night in the vicinity west of Shumway Lake, the third night at Charlie Humphrey's ranch near Knutsford. From there the riders got up with the dawn to reach town before too many people were on the streets, then to the Canadian Pacific Railway station and the corrals and weigh scales that Douglas Lake had built and still owned.

Early one summer morning, while the cowboys herded the irritable four- and five-year-old beeves to market, a housewife

84

was hanging her washing on the line. The flap and flutter of the white sheets in the still morning air scared the steers so much that they stampeded. It was several days before the cowboys managed to gather them to take them to the station.

If the shipping crew drove the cattle slowly and carefully, they would arrive in good shape at the weigh scales. They would fetch around 3½ cents a pound for steers, an average of $35 a head. Once the Kamloops station agent had weighed all the cattle, the cowboys loaded them, 20 to a boxcar, onto the Vancouver-bound train. Then the crew, usually with Joe Coutlee in charge, made the ride home in two days, often just in time to turn around and bring in the next drive.

Around 2,000 ranch-raised cattle and 700 purchased cattle went in this way to the coast market annually, initially to Porter's, with whom Greaves had done business for 25 years. In early 1909 Robert Porter and Sons sold their wholesale and retail butchering business—two stores in Victoria and two in Vancouver—to Pat Burns and Company. A good business relationship then developed between Pat Burns and Company and Douglas Lake Cattle Company.

Another person with whom Douglas Lake had long transacted business made news in 1901. That year, Johnny Chillihitzia, chief of the Okanagans at Spahomin since the death in 1884 of his father, Chief Chillihitzia, met the new King Edward VII. Along with two other local Indian chiefs and Father Le Jeune, Chief Johnny had an audience with the prime minister, Sir Wilfred Laurier, with King Edward, and with Pope Leo XIII in Rome shortly before the pontiff's death.

By the summer of 1902, Thomson was 75; Ward was 62; J.B. Greaves, manager, was 71. Greaves was no longer able to ride, as his horse had fallen on him while jumping an irrigation ditch. He now travelled around the ranch by horse and buggy. Greaves had hoped that his son, Peter, would take over the management of the ranch, but while in South Africa for the Boer War, Peter had liked that country so much that he spent the rest of his days there as a blacksmith.

So by the end of July, Thomson was receptive to the idea of selling the ranch to the Honourable Edgar Dewdney.

Thomson wrote to Greaves,

His idea is to form a company to take over the property.... When you have seen Dewdney if he means business—we will report to Ward. The following are figures I have given ... but have *not given him any price.*

Purebred cattle 500	@ $40.00	$ 20,000.00
2,500 steers	@ $45.00	112,500.00
10,000 (cows)	@ $25.00	250,000.00
1000 horses	@ $60.00	60,000.00
85,000 acres	@ $ 5.00	425,000.00
		867,000.00 [sic]

quantities to be verified.

Thomson was not willing to give an option on the place; "we have to be very careful not to get entangled." He was only prepared to allow time for Dewdney to make a reasonable offer. He never did.

Instead of selling out, C.W.R. Thomson wrote in January 1903 on behalf of Douglas Lake to the provincial government, offering to purchase the 16,000-acre Hamilton Commonage at $2.50 an acre. The B. C. Cattle Company, which owned the Triangle Ranch at Quilchena, made a similar offer. The government turned down both, however. All the ranches near the Hamilton Commonage, including Douglas Lake, continued to share it, running their cattle together.

By January 1903 Greaves could see that the winter was going to be a hard one and hay supplies were not sufficient, despite increased production; unless the ranch bought more, some of the herd would starve. After a lot of organization he arranged for 5,000 tons of hay to come by CPR from the coast to Mission Flats in Kamloops, and set out with 5,000 hungry cattle to the railhead there to feed. Driving such a large herd was difficult when drifting snow hid the deep ruts that thousands of Douglas Lake and other cattle had tramped over the years on their way to market, and in the white-out no treacherous gulleys or rocks could be seen.

A lead bunch of several hundred head of cattle broke trail, a crew of riders forcing them along. The rest of the drive strung out for miles, kept in line by riders on each flank and more on the "drag" who herded the tailenders. Footing was perilous for cattle, horses and men, and there was nowhere to stop and feed a herd of 5,000 along the way.

For four months, the 5,000 head stayed at Mission Flats feeding right out of the boxcars along the line of the railway track. Greaves's men fed the remainder of the Douglas Lake herd at Douglas Lake until 15 March when, with spring nowhere in sight and most of the ranch hay gone, the cowboys drove another 1,000 three-year-old steers 35 miles via the treacherous wagon road following the Salmon River canyon down to Grande Prairie. Each year Grande Prairie ranchers provided overnight holding fields for Douglas Lake's annual drive of cattle purchased from the Okanagan. In 1903 they provided 1,000 steers with winter feed of pea straw—the vegetation left after hogs had rooted through it for the peas; Grande Prairie was famous at this time for its peas and its pigs. The unorthodox forage saved the day, and Greaves arranged to take up to a tenth of the herd down to Grande Prairie for feeding each winter, a practice that was to last almost 70 years. Thomas Joseph Clemitson, known always as Tottie, became the annual overseer of this herd, buying sufficient hay from his neighbouring ranchers at Grande Prairie and from others at Falkland, moving the cattle from ranch to ranch, and supervising the feeding out of the hay and the care of the cattle.

In April 1903, the snows started receding around Douglas Lake, the earliest ranges began to green up, and the herd at Kamloops could return to the ranch. But most of the cows now had calves, and driving the herd was even more slow and tedious than the trip to the railhead. The cows, pushed onward by the cowboys, bunched up towards the front of the drive while the calves fell to the back, longing to return to where they last smelled their mothers and had a suck of milk. It was a painstakingly slow drive back to Douglas Lake, but at least the cattle were alive. Another winter was over.

In 1901, while the Boer War had been raging in South Africa, the Canadian government passed the South African War Land Grant Act. This entitled any Canadian who had been on active service in that war to a free land grant of 160 acres. Many young soldiers had no wish to go on the land, and eagerly accepted offers from $1.00 to $2.25 per acre for their negotiable scrip.

Like many another ranch with cash to spare, Douglas Lake spent $9,700 buying grants from 49 soldiers then living anywhere from Transvaal, South Africa, to Monte Creek, B.C. Almost 8,000 acres in various locations, from Kilroy Lake in the south to Cayuse Mountain in the north, enlarged the ranch boundaries in this way. Late in 1906, the Crown granted the deed for the last of this land to the ranch.

The death in 1904 of John Wilson, Greaves's former neighbour, set Greaves thinking. If he were to realize his plans for Douglas Lake before he became too old, he must push right on. Times were changing, too. Okanagan cattle ranchers were beginning to plant fruit trees, having recognized that their valley was excellent for growing fruit. Greaves made two trips in 1906 to buy up the remaining 4,000 head of cattle in the Okanagan Valley.

It was also one of the occasions when Greaves fired Coutlee for drinking and put another man in charge, this time Oliver Walker, a Colorado cowboy who had been working on the ranch for about a year. Coutlee went to work at the Triangle for Frank Jackson. During the summer when the cowboy camp was at the Hamilton, Greaves sent word for them to come in, but Walker, who had a girlfriend there, sent the crew in without him. Greaves fired him and hired Joe Coutlee back.

Up until this time, Douglas Lake cattle did not range farther south than Minnie Lake. Twenty miles to the south of the ranch was Aspen Grove—copper-mining country. The Portland Mining Company, now in its fifth year of operation, was considering building a smelter. But Aspen Grove was also an excellent summer range area, as the Bates boys and many others had recognized when they had begun preempting land there in 1904. Greaves decided to summer range his female stock to the south in this undeveloped timbered country in 1905. Within a short time he purchased land on the south end of Courtenay Lake and kept there two of the six purebred Clydesdale stallions the ranch then owned.

With the ranch cattle summering to the south, the Salmon River bottom lands to the north were free from cattle and could now be cleared of their dense brush so that more hay could be grown. But Greaves could not find the extra labour required, as he was already employing all the available work force from the Indian reserve and from his white neighbours. Someone suggested that he hire some "Canadians," meaning easterners. It was a novel idea, and Greaves thought he would try it. He brought in a dozen eastern Canadians, and put Tom Jones in charge of the Brush Camp at the north end of Fish Lake. Using axes, picks, mattocks, chains, horses and scrapers, Jones's crew cleared 155 acres. This cultivated area, added to the surrounding meadow land left in its native state, became the 400-acre Brush Field growing hay and oats. The eastern Canadians, as well as being good horsemen and axemen, could run a sawmill, and the various barns they built on the ranch lasted many years. Greaves was well pleased with his experiment.

In Victoria, the Victoria Gas Company sold out to its only competitor, the British Columbia Electric Railway Company, on 29 April 1905. Thomson received almost $4,000 worth of B. C. Electric Railway stocks, and over $100,000 cash. Flushed with the success of the gas company sale, he sent a civil engineer and surveyor, Dennis Harris, to Greaves two months later with a letter:

The bearer Mr. Dennis Harris has taken in hand the particulars of our property intending to try and sell the same. I have given him all particulars of my department and have attempted to make an inventory, putting the land at $7.50 p. acre and the cattle at $20 a head. Horses, I don't know. I have called the cattle 14000 head, allowing for surplus last year and your recent purchases. . . . My figures work out at about one million dollars. The acreage is 97,159. . . . I have told him the price is almost too heavy to effect a sale but he is willing to try.

Dennis Harris proved himself to be a most thorough man. By August he had drawn up a seven-page document giving details of the company. He valued the ranch at $1,105,427.50, of which just under half was for the land.

The cattle constituted one third. Of the 12,375 head, 5,000 were cows serviced by bulls at the rate of one bull for every 25 cows; 5,000 were steers; 2,200 were yearlings. The calf crop of 2,500 did not figure in the inventory, being too young to be of any value. If the herd were to remain constant, up to 2,500 animals—half older cows and half steers—could go to market each year. A marketable steer averaged around $44, a cow around $39, so that the annual return was over $100,000, while expenses averaged only $27,600.

The 498 horses were valued at $74,600, the six stallions making up $9,500 of that figure. Each year the 250 range brood mares dropped an average of 141 foals.

Machinery, improvements such as 300 miles of fence at $3,000, and trading stock in the store at $6,000 made up the balance of the valuation. Although the three shareholders were considering selling, they had authorized construction of a new store and bunkhouse, the old store not being sufficiently large for the increasing local trade. Observed Harris, "The bunk house besides giving additional accommodation will divide the White from the Indian labor."

On this topic, Harris stated,

The class of labor employed... Whites, Indian and Chinese... is easily obtainable at short notice and suitable to the work. [Some of] the average annual expense for labor... finds its way back in trade at the store with the exception of the Chinese, who are paid on a lower scale than the Whites. Cow Boys are paid... $25.00 per month and their board. Chinese $15 to $20 and the man in charge of the cattle market—$75 per month. A store-keeper, blacksmith, Chinese cook and assistants are continually employed the year round at the head station.

The average labour force numbered something over 100. Though the partners were allocating much of the profit to acquiring new land, they were taking a good share themselves: in 1901, Greaves, Thomson and Ward shared $99,728 in dividends. But this was a peak and in 1905 their profits were only $8,736. Nevertheless, they had put together a viable cattle ranch and it was no wonder that they were seeking to sell it, for, as Dennis Harris reported, "the combined ages of the three owners being over 210 years, calls for a rest from labor."

At the end of 1905, the shareholders ordered a compilation of total figures for the 21 years that their stock company had operated. These show that they had continued to buy cattle to run on other ranches until they were fit for market, but figures for that part of the operation are not separate from the Douglas Lake Ranch figures. Total gross income from horse and cattle sales since March 1884 had reached just over $3 million. Operating profits had purchased all cattle and land for a total price of $1.5 million. Operating expenses and bank interest had consumed $1 million. The four partners had together drawn over $500,000 from profits.

In April 1906 Thomson had cause to write again:

D. Harris is here. He claims to have sold to Western Land Co. and one of the Directors is on his way out—with Cecil Ward [a son of W.C. Ward]. . . . [This Director is] one of the irrigation engineers of the C.P.R. who with Ricardo of Vernon and Palmer of Victoria, the fruit inspectors, are to verify Harris' figures and frame a report. Meantime W.C. Ward has asked me to give him absolute authority to deal with the property. This I have done and I wait his instructions. The option expires on 12th May. After that date I will get consent of all partners before taking any steps. The question will come up of extending the time and after I have heard from Ward, I propose to make an appointment with Cecil Ward and his men, to meet them at D. Lake, go into their report and listen to any proposal.

The sale of Douglas Lake seemed imminent indeed.

CHAPTER NINE

In an address to the International Association of Chiefs of Police, at the Jamestown exposition, William A. Pinkerton gave credit to Old Bill Miner ... for first using the phrase 'Hands up!' while engaged in his professional activities as a highway robber.

Inland Sentinel,
6 December 1907

After centuries of irrigating paddy fields in Asia, the Chinese CPR construction workers could make a stream run exactly where they wanted it to, and their experience was therefore invaluable in flood-irrigating the hayfields of British Columbia. Each spring at Douglas Lake, these irrigators moved from camp to camp, repairing ditches and flumes and flood-irrigating as they went.

One day when George Edwards, a kindly old fellow working for Greaves on the Home Ranch, was taking some Chinese irrigators by two-horse wagon to another part of the ranch, something scared the young horse of the pair and it shied, pulling the wagon off the trail into a rough meadow. The wagon bounced and jolted so much that a workman fell out, fatally hitting his head on a rock. The cook at the Home Ranch, a brother of the dead irrigator, was furious with George Edwards for being so careless in handling the team, and threatened to poison him. J. B. Greaves was so concerned for Edwards's safety that he let him go. Edwards returned to prior haunts around Aspen Grove and Princeton, staying with Jack Budd, a rancher.

George Edwards's eyes were a piercing blue, his nose long and straight, his hair grey and wispy, his moustache distinguished, and his accent Texan. He was a most likeable and generous man who gave youngsters Sunday rides on his well-trained horse, Pat, flooded a field in winter for a lonely little girl to skate on, and always had gifts of candy for both the children he saw and the adults he visited. Being a shoemaker, he cobbled many pairs of shoes for children who might otherwise have gone barefoot, and he paid for the education of many children as well as teaching them himself in Sunday School. He enlivened social functions: being light on his feet, he was an excellent dancing partner; he could make a fiddle sing; and he could bring tears to the eyes of his listeners with his rendering of "My Old Kentucky Home."

In mid-March 1906, George Edwards and two of his friends, Shorty Dunn and Louis Colquhoun, went prospecting on the South Thompson River. Dunn was an experienced prospector, his most recent employment being at the Ashnola mine near Hedley. Colquhoun's background was varied: he had taught school near his home in Ontario; worked on a survey crew in Calgary; kept books elsewhere; prospected at Phoenix in the Kootenay mining district.

Their route took them through the good prospecting country on the easterly border of Douglas Lake Ranch. After six weeks' travelling they reached the South Thompson River, east of Kamloops. They bought more provisions from one of the local ranchers, Albert Duck, and camped in the area, setting out each day to learn the country.

On 8 May, at 11:30 p.m., the Vancouver-bound CPR Imperial Limited Express No. 97 was just half a mile west of the small station at Ducks, next stop Kamloops, when engineer Joseph Callin spotted two men on the water tender, creeping over the coals towards the engine cab. Thinking it was a joke, Callin called out, "What the mischief are you doing there?" His laughing stopped when the taller of the two masked men, wearing driving goggles and a black handkerchief to hide his face, pointed a large pistol at his head and said, "We are going to rob this train." The second masked man covered Callin and fireman Ratcliffe with his gun while his partner searched them for weapons.

"I want you to stop the train at the 116 Mile post and when we stop, we will cut the coach off," said the leader. Callin did as he was told. At gunpoint, Ratcliffe detached the engine and mail car from the rest of the train. A third man came running across the field, carrying a large parcel. The sweater he held to his nose was not sufficient disguise to prevent Callin from having a good look at his face.

Aboard the engine again the robbers commanded Callin to pull ahead to a flume, where the mail coach door was opened. "Hand over the registered mail," ordered the tall man with the black mask and goggles. The mail clerk pointed out the bags. "Is this all? Where's the bunch for 'Frisco?" the robber demanded. The clerk explained that this was the mail car, not the express car. In his anger at being foiled, the robber allowed his mask to slip for an instant, giving the clerk a good look at the amber grey moustache and aging face.

The robber pocketed the four pieces of registered mail and ordered Callin, "Run ahead a mile and a half." Between mileposts 119 and 120 the three robbers got off the train, calling politely as they disappeared into the darkness, "Goodnight, boys, and take care of yourselves."

It was believed that the train robbers had been after a big shipment of relief money going to the victims of the San Francisco earthquake, and they did miss this shipment by just a day or so. A rumour circulated that they had got away with less than $20 and the catch phrase of the day became, "but the CPR robs everyone every day."

Urgent cables from Kamloops soon brought CPR detectives and superintendents. Men from the Pinkerton Detective Agency and Thiel Detective Agency of Seattle rushed to the scene; Superintendent F.S. Hussey of the Provincial Police Force took charge. A reward notice for $11,500 brought many more would-be detectives rushing to the aid of law and order.

At daylight an Indian tracker came upon a campfire near the railway. The coals were still warm and in the soil around them the Indian easily distinguished the tracks of two pairs of hobnailed boots and one smooth-soled pair. The robbers' trail was established, and the chase began.

The Indians were splendid trackers and hunters, and the constable in charge of them, William Fernie, had received invaluable schooling five years before when he had been searching for Boer guerillas in South Africa.

Losing and finding the trail time and again, the small posse travelled south, where it was joined by Constable Pearse and more men. Later in the day they discovered the camp that the robbers had used for planning the holdup. Situated on a big hill six miles from Ducks, this secluded site had afforded a safe hideaway and an easy route to the railway. The posse centred its activity around the local landmark, quickly titled the Big Hill, where Pearse found two hobbled horses that he believed were the robbers' getaway horses. The mail thieves were now afoot. Then a cache of riding and camping equipment turned up, and also the charred remains of some of the opened mail. Pearse sent a messenger to Kamloops for bloodhounds to help the Indian trackers follow the robbers.

On Friday the eleventh, new tracks were found leading south through a timbered valley, but everything seemed against the posse's being able to follow the trail, for a drenching rain that turned to snow the next night threatened to blot out the tracks. By Sunday, five days after the robbery, the posse had travelled many miles, but had been sidetracked so often following trappers' trails and short cuts through swamps that it had actually covered less than 20 miles of the robbers' route.

Fernie was getting impatient. He decided to follow up a hunch. The tracks were leading the posse consistently south towards Chapperon Lake. The day before, there had been a report that unknown tracks had been spotted in Batchelor's Meadow, Douglas Lake Cattle Company's most northerly haying area, which lay directly south of their present position. Fernie thought it likely that the wanted men would continue their southerly route and would follow the wagon road from Chapperon Lake through Douglas Lake Home Ranch. Pearse was determined not to move farther ahead until the bloodhounds arrived, so Fernie set off alone on horseback.

After a long day's ride, he camped Sunday night at the Brush Camp, where he heard from the foreman, Tom Jones, that someone—Fernie was certain it was one of the robbers—had come to the cabin door the night before, then disappeared into the darkness. Though Jones's crew had immediately searched the area, they found no one.

The next morning, wishing to travel light, Fernie left his

carbine and shoulder strap behind, a decision he was to rue later in the day. He struck across country, heading southwest towards Douglas Lake Home Ranch. He saw nothing unusual along the 15-mile route, but from J.B. Greaves he heard that the Calgary detachment of the North West Mounted Police had stayed at the Home Ranch overnight and were now heading for Chapperon Lake to await Pearse. There were seven of them, all under Acting Sergeant John J. Wilson.

Disappointed that his hunch had not paid off, Fernie decided to catch up with the Mounties, turned his tired horse around, and set out to retrace his steps. Less than a quarter of the distance there, near Murray Creek, Fernie spotted three men heading towards him, walking abreast. As he got closer he recognized the three from descriptions of the robbers.

Fernie was jubilant, yet at the same time helpless without his firearm. It was already too late to turn back, and the young constable realized that he could not arrest the three on his own. He decided to act like any other traveller, and plodded along on his horse as though he had not a care in the world.

"Hello, there," called the oldest of the three men. "Which way is Quilchena?"

Fernie gave directions and asked in turn, "Am I on the right road for Chapperon?"

While a casual conversation continued, Fernie saw that the youngest man kept one hand in his coat pocket all the while, undoubtedly on his pistol. A revolver protruded from the third man's pocket. Urging his horse forward, Fernie bade the three farewell and made as if to continue his journey.

Eventually, he looked back and watched the men walking in the distance. He waited until they were out of sight and then rode back to where he had met them. There were the three telltale tracks, two pairs of hobnailed boots and one smooth-soled pair.

Taking great care to avoid their path, Fernie galloped his horse back to the Home Ranch. J.B. Greaves loaned him a fresh horse and a gun, and minutes later Fernie was again heading for Chapperon Lake, racing as hard as the horse could go.

When he arrived, the Mounted Police had just finished their dinner and had unsaddled for the day. They were awaiting Pearse and the posse. As the policemen saddled up once more, Fernie hastily told them the story, and within three minutes they had set out.

At one o'clock, they were at the spot where the constable had last seen the robbers, having ridden the six miles in less than 20 minutes. There was no sign of the wanted men. For the third time that day, Fernie headed for Douglas Lake Home Ranch to send a messenger to Pearse, and another to Quilchena for more bloodhounds.

The police spread out to search the thick brush. Suddenly Corporal Stewart waved his Stetson to the others. He had seen a thin wisp of smoke ahead. The Mounties rode towards the smoke and found three men sitting around a campfire, finishing their lunch.

"Where have you come from?" Stewart asked. The leader, as cool as ever, replied, "Grande Prairie. We've been prospecting."

"I guess you're the men we want for holding up the train," said Sergeant Wilson, covering the leader with his revolver.

"We don't look much like train robbers," denied the leader calmly. But he had barely finished speaking when the shortest of the three jumped to his feet and ran away, calling back, "Look out, boys, it's all off." Corporal Brown headed after him.

Sergeant Wilson continued to cover the leader. Corporal Peters covered the other man left on the ground just as he made a grab for the revolver lying beside him. Peters threatened to blow out his brains if he made another move, and grabbed the revolver himself.

The runaway robber swung around and fired at Corporal Brown, who shot back. Corporal Peters fired also. Five or six shots volleyed back and forth, then the robber fell into a ditch, calling out, "It's all up. I'm done for."

The police went up to arrest him, and found he had suffered a leg wound. His revolver was empty, but he was still defiant, boasting, "If I had had my automatic revolver, there would have been a hot time around here, you can bet."

While searching the robbers, Sergeant Shoebottom noticed

a tattoo on the thumb base of the leader and cried out, "Look out for the old fellow; that's Bill Miner!"

Bill Miner was a household name across North America. From the age of 16 he had been successfully holding up and robbing stagecoaches and trains. Three times he had been imprisoned in San Quentin jail, and twice he had escaped from that hell-hole prison. By 1901, at the age of 54, he had spent almost 33 years behind bars, all in San Quentin. Escaping capture after a train robbery in Oregon in 1903, he had disappeared until now.

Transporting the robbers by a team and wagon brought out from the Home Ranch, the escort stopped first at Douglas Lake headquarters. J.B. Greaves and many of his crew came out to meet Fernie and the North West Mounted Police, and, astonishingly, shook hands fondly with Bill Miner. Greaves turned in amazement to the police and informed them that they had made a terrible mistake, for this was no train robber, this was George Edwards, the kindly and respected gentleman from Aspen Grove.

But George Edwards and his two friends, Shorty Dunn and Louis Colquhoun, were the three who had robbed the train west of Ducks. And George Edwards was just one of the many aliases that Bill Miner had used before, and would use again. In fact, many years later, it became known that even the name Bill Miner was an alias. George Edwards and Jack Budd, the man Edwards had stayed with in Princeton, were brothers, and their surname was McDonald.

In Kamloops, Miner and Dunn were sentenced to life, and Colquhoun to 25 years, a stiff sentence for a novice. They were to serve their terms at the New Westminster federal penitentiary.

Colquhoun died in prison. Dunn's leg wound healed, and in 1918 he was paroled; he later drowned saving a man's life. On 8 August 1907, just 13 months after entering the penitentiary, Bill Miner escaped. Despite being recognized twice in British Columbia, he evaded recapture.

Some three years later, Miner was caught staging the first train robbery in the state of Georgia and was sentenced to imprisonment in the Georgia State prison farm. He died there three years later on 2 September 1913.

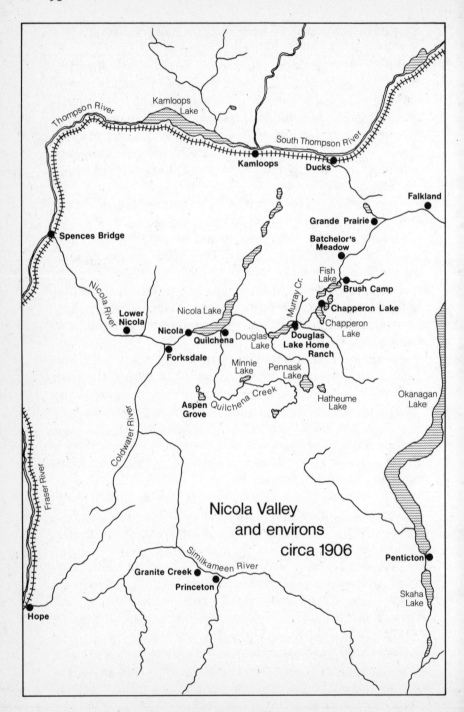

Nicola Valley
and environs
circa 1906

CHAPTER TEN

I have been bothered considerable this spring and summer with your cattle. It seems that you must have had someone pulling our fences down right along so as to let your stock into our fields. I hope you will not force us to take extreme measures to protect ourselves. We will let you have the use of the field for any reasonable figure so as to avoid further trouble.

Goodwin Bros. to J.B. Greaves,
5 July 1905

The Nicola Valley was changing: what had been purely ranching country since 1868 was about to become a mining centre. The catalyst? The long-expected CPR branch line from Spences Bridge to the Nicola coal basin, to be built at a cost of $1 million. By the summer of 1906, freight was travelling along the finished track from Spences Bridge to Coutlee. The Bank of Montreal was opening a branch at Nicola and many speculated that several hotels, like the new Quilchena Hotel which Joseph Guichon was building, would spring up throughout the valley to accommodate the tourists from Vancouver.

As Nicola waited for the railway line, J.B. Greaves awaited the arrival of the Western Land Company directors who held an extended option on the ranch. On 5 July, three representatives arrived—R.M. Palmer, fruit inspector for B.C.; W. Crawley Ricardo, manager of Lord Aberdeen's Coldstream Estate in Vernon; and Albert E. Ashcroft, irrigation engineer for the Coldstream. For 11 days, they studied the ranch operation and the possibility of doubling the irrigated hayland acreage economically. Extra hay would make the purchase of the ranch even more attractive.

Cecil Ward and two others took over from the Western Land Company by forming the Dolaca Syndicate, a group interested in promoting a sale of Douglas Lake Cattle Company. In February 1907 this syndicate offered Greaves,

Thomson and Ward £200,000 for the ranch; £185,500 in cash, the rest in ordinary shares of Canadian Estates, Limited. Dolaca Syndicate, for a fee of £70,000 in shares, was acting on behalf of Canadian Estates, a public company interested in acquiring Douglas Lake. The ranch shareholders agreed to sell.

Canadian Estates, Limited opened their subscription list on the London stock exchange on 11 March. The prospectus advertising the new company was a 16-page glossy edition complete with photographs, maps and reports of the cattle ranch at Douglas Lake. The *Inland Sentinel* was impressed by the distinguished directors—among whom were Cecil Ward, who had practised law in Kamloops, and Rt. Hon. H.O. Arnold-Foster, M.P., formerly war minister for Great Britain.

Though complex, the sale seemed to be advancing satisfactorily when suddenly Greaves received a note from Thomson. "They [Canadian Estates] have tried to float a company to get off shares to amount of £400,000, say two million dollars and have failed—and they now want us to take shares in part payment for the D.L. Property. . . . My terms are $900,000 net. I would not take a single share so that any one could say I had been connected in deceiving the public." W.C. Ward was in Victoria, eager to see Greaves and persuade him to go along with the new plan, but Thomson was adamant. "When we do meet Ward, [we] can make any proposition as to future management of the Company, but the idea of taking shares and coming under a management of a Board of Directors in London, all provided with salaries, traveling expenses, it is too absurd, and while this is going on the Promoters who get their shares for nothing will be quietly selling out and pocketing the cash."

Thomson was ready to take the property off the market, Greaves was with him, and it was not long before Ward agreed.

Though there was no change in ownership at Douglas Lake, the trains that had been running from Spences Bridge to Nicola since 28 March 1907 were bringing about enormous changes in the area. The four ranches belonging to Jesus Garcia, William Voght, and the estates of John and William

Charters constituted Forksdale, the small settlement around the confluence of the Nicola and Coldwater rivers, and they were sitting on some of the best coal beds in the Nicola Valley. Many individual and corporate speculators took up coal rights and options to purchase for this whole area: soon a new town began to mushroom. It became known as Merritt, after Colonel William Hamilton Merritt, the railway contractor who had kept his supply camp at Forksdale.

Until now, the town of Nicola had been the commercial, industrial, agricultural and ecclesiastical hub of the Nicola Valley, but with the completion of the railway, Merritt took over as the commercial and industrial centre. Nicola became even more important to the ranchers, for now cattle, sheep and horses could be shipped from the new railhead that was so many miles closer than Kamloops. Douglas Lake Ranch beef cattle were now driven a much less tedious route to Nicola instead of to Kamloops. The ranch grazing patterns changed accordingly and the Gilmore Field, which lay on the route to the new station, came into greater use.

In 1891 Douglas Lake acquired as a neighbour the baker Alfred Goodwin. Al and his brother Fred preempted land north of Fish Lake not far from Owen Batchelor's meadow, their headquarters being the Upper Place. When the government threw open the 29-square-mile Marsh Meadow Commonage the same year, Al acquired three half sections; one personally, one through Fred, and one through Ned Williams, a Cornish copper miner with whom Goodwin had a sharecrop arrangement.

This new land ran north from Rush Lake to the west of land owned and grazed by Douglas Lake, and a line fence was needed to keep the two herds separate. Greaves and Goodwin agreed to share the cost equally, with Goodwin arranging for the fence to be built by contract.

Goodwin gave the fencing contract to James Madden, a well-muscled man of 240 pounds with wild sandy hair, nicknamed the Big Kid. He was well known for his unusual domestic habits, for he lived happily in a tiny, rude shack hardly big enough to crawl into. Once, he contracted to make charcoal to fuel the blacksmith's shop at Douglas Lake, and to be close to work built himself a shack in the Boar's Nest. Most people believed that the field was so named because of Madden's messy way of living.

Madden was working on the line fence one day when Greaves rode by and, in conversation, mentioned the price of the contract. Money meant little to a man with such few wants, but the Big Kid corrected Greaves and said that he was being paid a lot less by Goodwin. Goodwin was overcharging Greaves for his share of the cost of the fence. An angry Greaves immediately confronted Goodwin and informed him that he would build his own line fence. And so, two parallel fences, separated by a few feet, crossed miles of rangeland, a monument to the greed of Goodwin and the pride and stubbornness of Greaves. It was also a daily reminder of the enmity between the two ranchers.

This double or "mad" fence defined the eastern boundary of Goodwin's new land. His western boundary fence took in half as much land again as he rightly owned. Then he built a third fence farther west taking in a very large portion of Crown land that had been open grazing for Douglas Lake cattle. The ambitious Goodwin "was determined to get all the pasture he could; he could always hope to get the stock to make use of it." The animosity between Greaves and Goodwin deepened.

Some of Al Goodwin's relations came out to help him preempt a bigger spread, although none stayed long. Goodwin himself took a trip back to Norfolkshire, England, got married, and returned with his bride and with some young men who had agreed to work for him for one year. One of them, Talbot Bond, later said that Goodwin never "got so much work done for so little money before or after."

The Upper Place was no longer headquarters for the Goodwin spread. Rush Lake, later called Sawmill Lake, was more central, and here Goodwin brought in machinery to cut the lumber for his main ranch house. The Goodwins called their ranch "Norfolk"—a name that stuck.

During 1904, the fences Goodwin had built around the Crown land, which he now called his Brush Field, were constantly being broken down, and he understandably believed that Greaves was responsible. Early one morning, Ben Goodwin, Al's brother, saw some men riding away from a hole in the fence and fired his .30–30 over their heads. For a while no more fences were broken.

The attack on Goodwin's fences began again in 1905, and he wrote to Greaves, "You ordered your men to turn Horses & Cattle into my field. I put the horses and some of the cattle back again. We shall have two men herding that field. If you would make me a reasonable offer for the use of the pasture it would save both you and me a lot of trouble." Naturally, Greaves was not prepared to pay for the use of the Crown land that Goodwin had simply fenced in.

Goodwin was on better terms with his neighbour to the west, Billy Lauder, who had purchased his father's Springbank Ranch in 1896. In 1906 Goodwin sold Lauder all his beef cattle to fill an order and at the same time tried, unsuccessfully, to sell him a milk cow. Lauder declined, for he could find only the brand SS instead of Goodwin's 77 or a vent to prove new ownership. That was the summer when Greaves hired and fired Oliver Walker as cowboss at Douglas Lake. Feeling somewhat resentful, the American cowboy hired on with Goodwin as his principal range rider.

Lauder arranged to buy Goodwin's beef steers in 1907 also. On 21 August, he rode with a hired man to the Upper Place to collect the cattle that Goodwin advised were grazing in his Marsh Meadow Field, southwest of his Brush Field. The two ranchers agreed to meet there around four o'clock that afternoon and Goodwin began giving directions on how to get to his field, but Lauder knew that Goodwin was describing a long way round and determined to go through the Brush Field.

In the southwest corner of the Brush Field, Lauder came across a bunch of 25 horses. They were handsome two-year-olds, mostly Clydes, their tails docked. The rancher immediately recognized them as Douglas Lake horses, yet over half of them had their brand, SS on the right shoulder, blotched as badly as the brand ||| on the milk cow he had again seen earlier that day at the Upper Place. Lauder wondered at Goodwin's stupidity: trying to sell him a milk cow he did not own and disfiguring the brands of Douglas Lake horses. Both were indictable offences.

It was six o'clock when Goodwin joined Lauder at Marsh Meadow Field and Lauder began cutting out the beef he needed. Soon he had the job done, but Goodwin pointed out that by contract Lauder had to buy all the cattle that were for sale. Goodwin was particularly insistent that Lauder buy a bunch of three-year-olds, but Lauder insisted that they were only two-year-olds and would only pay a two-year-old's price.

Goodwin snarled, "You little son of a bitch."

"You're another," retorted Lauder. "I've seen enough today to put you in the chain gang!"

Goodwin's face flushed. "Get off your horse and they'll put you in a coffin!"

Lauder did not wait to find out if Goodwin would carry out his threat, and as he rode away was sure that a gun was pointing at his back.

In the past, Greaves had loaned Lauder money. Lauder had reciprocated by buying hay for Douglas Lake, for he could buy cheaper than Greaves (the mere sight of whom could send prices up) and Greaves paid him a commission: all the hay he needed to winter his own cattle. Lauder headed for Greaves and the Douglas Lake Home Ranch.

Two days later, Coutlee and Whiteford went to search for the horses Lauder had seen. Near the small lakes in Goodwin's Brush Field they found 23 horses, 13 of them with their brands disfigured. They drove the animals to the Home Ranch, and Greaves sent to Nicola for Constable Walter Clark. The next day, the constable and Whiteford searched further and located two more horses.

Although obviously a crime had been committed, there was little evidence to lay charges against Goodwin. Douglas Lake horses had been grazing in Goodwin's field, but no one had seen them driven there, nor had anyone seen who had disfigured the brands.

Greaves wrote two letters: the first went to the *Kamloops Standard* offering a reward of $1,500 to anyone who could help in the arrest and conviction of the horse thief; the second went to Thomson, explaining the situation and asking him to seek a further reward from the government. In October, Thomson replied, "I send you enclosed the Government offer of $500. It is the best I could get them to do." A reward of $2,000, however, was enough to bring results.

That fall, Goodwin stopped selling his cattle to Lauder, and instead of shipping through Kamloops or the popular new railhead at Nicola, he had Walker drive his beef to the quiet station at Ducks.

On 25 January, Walker left Goodwin and headed for Kamloops to take the first train east. He deposited his baggage at the station and, having some time to kill, wandered through the town, where he met Jack Whiteford. As a result of their conversation, Walker collected his baggage and went to Douglas Lake with Whiteford. After talking with Greaves, the two headed for the attorney general's office in Victoria.

On 6 February, Whiteford appeared in front of Constable Earnest Pearse in Kamloops to lay the information and complaint against Alfred R. Goodwin. He charged that Goodwin "did fraudulently take hold and keep in his possession... fourteen two year old horses the property of the Douglas Lake Cattle Co.... and... did fraudulently obliterate and deface the brand of the said Douglas Lake Cattle Co. on the said fourteen two year old horses...."

Constable William Fernie arrived at Sawmill Lake the next day with a warrant for Goodwin's arrest, plus a search warrant to look for various medicines that Walker had said were used to change the brands. The obliteration of the brands on the 14 horses was not the only charge against Goodwin, and Fernie, Whiteford and Joe Coutlee accompanied Oliver Walker to collect more evidence.

At Goodwin's Upper Place, Walker found the hole where he and Goodwin had buried the hide of a Douglas Lake steer after they had shot the animal for meat for the haying crew. Coyote tracks led from the empty hole. Walker next led the group to some of Goodwin's weaning pens and brought out a young unbranded calf, and the group watched as it gambolled to the side of its mother, who was clearly branded with JL, Billy Lauder's brand. Later, Lauder identified both the cow and calf as his.

Nine separate charges were brought against Goodwin, and the preliminary hearing began on 18 February in Kamloops. Oliver Walker's testimony was most damaging for Goodwin. On 12 July 1907, according to Walker, he and Al Goodwin had ridden into Greaves's Marsh Meadow Field and rounded up 25 horses. They had thrown 14 of the 25 in the Marsh Meadow Corrals, now called the Horse Thief Corrals. They had then treated their brands, he testified, with either Fleming's Lump Jaw Cure or Fleming's Spavin Cure. These medications caused a blister that would slough from the hide, taking with it the lump or spavin, and they would remove a brand in the same way. Goodwin, said Walker, planned to ship these Douglas Lake horses once the resulting scars had healed.

On the advice of counsel, A. D. Macintyre, Goodwin reserved his defence, and was committed for trial at the Vernon spring assizes. As evidence for the charges accumulated, some of Goodwin's more unorthodox methods of ranching came to light. He raised large numbers of pigs which mostly fended for themselves, but occasionally he would "kill any cattle handy and cook the meat in a large boiler for them." George Nichols, Goodwin's carpenter in 1904, remembered his cooking cowhides for them also. The summer haying crews would go through a beef each month, and several neighbours thought that Goodwin "did not examine the brand too closely when he wanted beef for the larder." There were rumours that Goodwin had helped himself to grain out of the Chapperon yards. And there were other citizens who simply referred to Goodwin as "a crooked bugger."

Goodwin began to plan for a stay in jail and asked two young men, Talbot Bond and Bob Moir, to rent the Norfolk Ranch for a five-year term, with an option to purchase at a set price within ten years.

To Mr. Justice Irving at the Vernon spring assizes, Walker testified that he and Goodwin had experimented with Fleming medications on one of Goodwin's mares and on a colt. But A.D. Macintyre cleverly cross-examined him and every other witness, making everyone—especially Walker—appear crooked. He tried to show that across the border Oliver

Walker was known as a cattle and horse thief who had learned from a Colorado veterinary surgeon, Dr. Solont, the treatment for spavin. This had given him the idea for rustling Douglas Lake horses. He argued that Walker's motive in turning King's evidence was not to see justice done but that he was just greedy for the reward money. Ben and Mrs. Al Goodwin both incriminated Walker, and when Alfred Goodwin took the stand he denied completely any involvement in tampering with the brands of the Douglas Lake horses.

The jury members deliberated for many hours without reaching agreement, seven being for acquittal and five for conviction. Mr. Justice Irving held the case over until the fall assizes, and released Goodwin on bail for $7,000.

On 5 October, the new trial began in Kamloops in front of a new judge and jury. From Wednesday until the following Tuesday the jury listened to the evidence; then they retired for less than two hours and returned with a verdict of "Not Guilty."

Oliver Walker went to work for Greaves at Douglas Lake, on the premise that it was better that he work for the company than against it. Talbot Bond and Bob Moir purchased the Norfolk Ranch from Goodwin in 1912, improving the operation each year. Al Goodwin and his wife moved to Monte Creek, where they ran a few horses, kept an apple orchard, and made apple cider. They lived there quietly until her death in 1929 and his in 1936.

Getting rid of such a bad neighbour was a relief for Greaves, but it did not help him with his major problem—his age. To run one of the biggest outfits in the country was hard work for a man aged 77, even for Joseph Blackbourne Greaves.

CHAPTER ELEVEN

*I have just written to J.B.G. telling him that I have instructed Frank
to arrange a meeting with him & if possible also with you in order
that the question of Frank's services being secured by the D.L. Co.
may be settled definitely. . . . we have grown, & are growing older, and
there is no use in mincing matters.*

W. C. Ward to C.W.R. Thomson,
9 March 1909

Frank is not the man for the place.

C.W.R. Thomson to J. B. Greaves,
29 March 1909

The Nicola Valley had altered greatly in the quarter century
since Beak, Greaves, Thomson and Ward had formed
Douglas Lake Cattle Company. Transportation had so
improved that ranchers no longer had to drive their cattle to
Hope via the Coquahalla Pass and then ship them by steamer
to the coast. Now the cattle travelled to market by boxcars
from stations at Kamloops, Merritt and Nicola.

Automobiles had appeared, although ranchers accustomed
to buggy and horse were skeptical of an internal combustion
engine that seemed to start or stop on a whim. The first car in
Merritt was brought in by W.H. Armstrong, managing director of the Nicola Valley Coal & Coke Company Limited, in
1906. In fact, Armstrong had brought the first automobile
into British Columbia—a Stanley Steamer that he had
purchased in Boston in 1899. A year later Thomson told
Greaves that Ward had gone all the way from Victoria to Al-
berni "in an *Auto*."

An English syndicate announced in June 1908 that it had
taken an option on the ranches of James Aird, Robert Scott
and John N. Moore, which lay close to Nicola Lake, and the
ranch of Joseph Collett farther south. The syndicate planned

to subdivide the ranch lands into small holdings for orchards. They mapped out a full-sized town on the extensive Beaver Ranch land which they bought from young John N. Moore, who had inherited it from his father, Samuel, in 1900. No doubt the speculators had great hopes of selling the small parcels on the English market, for they also bought Scott's ranch.

When the option on Joseph Collett's 6,000-acre ranch lapsed, C.W.R. Thomson considered buying it, for being in a solid block east of Courtenay Lake, the Collett fields were en route to summer range and would therefore make useful pasture for those cattle being turned out to the Aspen Grove country, or as a holding area. "If Joe Collett's land is fenced, it is [a] good bargain at $4 per acre and I will join you," wrote Thomson to Greaves. "Ward as you know won't put any more money in and we will have to charge D.L. Co. rent or interest."

In February 1909, Thomson had outlined the deal. "The proposition is 6000 acres at $4 per acre, 3000 down, balance on receipt of crown grants. I should prefer to take conveyance at once of the 3000 acres, as life is uncertain and sometimes short. Anyway you can complete the deal."

It was no wonder that the uncertainty of life concerned Thomson, for he was 81 and had been in poor health for some years. W.C. Ward, the junior of the trio at 66, was resting in the sun on the Mediterranean coast. But he was not completely idle, for he had just written to Thomson and was anxiously awaiting a reply. Thomson forwarded Ward's letter to Greaves. It concerned Ward's son, Frank.

"I should be very glad if Frank could be taken on at Douglas Lake, under J.B.G.'s management & control as a second in Command. . . . Frank has had a long experience of Ranch life now and his wife is quite happy & helpful in it, so there is no question in *my* mind that we *ought not* to let him tie himself up in another direction but make this opportunity our own as well as his."

Thomson's attached note to Greaves showed that he had been anticipating this request. "He has always been hankering to get Frank taken on again. . . . I have always understood that you looked on Frank as not suited to the business." He advised Douglas Lake's manager to tell Ward "that you have made your arrangements with Joe Payne and that you cannot disturb your plans."

Francis Bulkley Ward had been born in Victoria, in February 1875, the sixth of Lydia and William C. Ward's ten children. Governesses had provided his early education and he planned to enter the Royal Military College, Kingston, Ontario, but a serious illness when he was 16 destroyed any chance of a military career. He recuperated on the Chilcotin ranch of his uncle, Mortimer Drummond, where he learned to throw a diamond hitch, pack ponies, and brand cattle.

Still 16, he entered the Bank of British Columbia in Victoria as an apprentice for three years, but he was dogged with ill health which left him with a chronic chest condition. The doctor advised Frank to find work outdoors, and he joined the Douglas Lake crew as a cowpuncher in November 1898, at $30 a month plus board.

J.B. Greaves and Joe Coutlee merely suffered the mild-mannered young man of 23 who shaved every day, seldom swore, refused their alcohol at 5:30 in the morning, and rode English in riding breeches and leggings on a flat saddle, jumping every ditch and stackyard fence in his path. The two cattlemen believed him to be unsuited to the life of a rancher: his strange "English" customs proved him so.

He left the ranch at the end of 1899 and, after working in the bank again, joined Joseph Pemberton, son of his father's friend, J.D. Pemberton, in buying the 44,000-acre "Two Dot" Ranch on Mosquito Creek, west of Nanton, in the foothills of southern Alberta. In January 1903, at the age of 28, Frank resigned from the bank to run the joint cattle venture. Three years later he married Ethel Frances Kennedy, daughter of a NWMP surgeon. Kenny, as her friends knew her, was an excellent horsewoman.

Horses were Frank's passion and he could take an Indian cayuse, break it, ride it, and train it as a polo pony. (In horse trading he obliged the Indians by using U.S. silver dollars as tender, laying the coins in a line on the ground so that they could be counted.) As captain of the Calgary polo team from 1907 he took a polo team to California each year and found a ready and profitable market for his trained ponies there.

But the grazing days of the prairies were disappearing as more and more grain growers surrounded the Two Dot, and the Alberta government became unwilling to lease large acreages for cattle. There was insufficient grass on the ranch to fatten the number of cattle with which he and Joe had begun, and not enough profit to support them both. Joe, surveying in British Columbia, became willing to buy or sell.

W.C. Ward was pressing his partners to take on Frank as assistant manager to Greaves.

I regard it as *absolutely essential* to the future welfare of the Co. that there should be somebody who can command the confidence of all three of us to take up J.B.G.'s work & responsibilities. . . . What could be more appropriate than call J.B.G. President, & appoint the son of one of the partners as his assistant manager. . . . Frank is to my mind especially fitted to fill the bill—absolutely honest & straight, steady & hard working as they are made. Very capable with cattle & having a knowledge of the D.L. property, all this & having carried on Ranch business now for years. I think Frank's services should be engaged at say not under $150 per month & found. . . . I know that J.B.G. has worked for that himself & would consider it perhaps high! But what is it when you consider how completely you are in the hands of your manager? & Frank would earn it & *more*. . . . I say we must provide for the faithful and competent management of the Co. being continued, & we can only do so satisfactorily by having a man with a distinct *personal* interest in the Co. . . . —a few hundred pounds a year for this is a trifle not worth considering in the balance!

. . . don't forget that we are each *equally* interested in following the right step & that *my* opinion & recommendation, ought not to be without full weight, especially as I originally initiated this Co!

But Thomson lacked such confidence. "If F.W. can't run a place for himself, I for one don't want any of his management." In another letter to Greaves, "Best say at once if we must have a manager, Frank is not the man for the place, but if W.C.W. is settled upon forcing him in the best way out of the tangle is to settle a price, buy or sell." And again, "How would it be for Frank to buy out Lauder and work his own ranch."

Thomson found another buyer in August 1909—Patrick Burns. The ranch value had reached $1,250,000, but Burns "values our property in an offhand way at $810,000. Would prefer to purchase the livestock and lease the land. Says there is too much dry land. Driving over the wagon road would give no idea of the back country. . . . Burns wants to get the property and says he will spend a time with you and go over the land."

Burns's plan was to buy the cattle and irrigable land and leave the rest. He had not become a superlative rancher and meat merchandiser by purchasing anything he considered unnecessary. But the shareholders were proud of their ranch and did not wish to see Burns cut it up. They turned down his offer.

On 27 October 1909, an unsigned agreement materialized which stated in part that "the property of the Co. may be sold at the price of Nine hundred thousand dollars net for cash or its satisfactory equivalent."

But far more intriguing were the next paragraphs:

[It is agreed] that the services of Frank B. Ward be engaged as Assistant to the Manager of the Co.'s business at Douglas Lake for one year from 1st November 1909 at a salary of One hundred & fifty dollars per month with board and lodging free, such engagement to be terminable on 1st Nov 1910 & thereinafter upon six months notice in writing. . . .

It is understood that the management of the affairs of the Co. remain in the sole control of J. B. Greaves as long as he remains as Manager & F. B. Ward will in all respects take his instructions from him & work entirely under his orders & advice.

This was quite a turnaround by Greaves and Thomson. They had been worn down by Ward's insistence. Frank dissolved his partnership with Joe Pemberton and bringing his wife and daughter, Betty, then almost two, moved to Nicola. He commuted on horseback to the Douglas Lake Home Ranch and his job under Greaves each week. To Greaves's chagrin, Frank even shared his living space in the Ranch House.

Frank soon began to object to the distance he had to ride to work and asked Greaves to have a house built for his family at the Home Ranch. Thomson advised Greaves, "It is not to our interest to put up a house, unless you require a house for general travellers' accom[m]odation." And later: "I see that Frank is punctual in his drawings [of salary]. Hope you get work out of him."

W.C. Ward found the next prospective buyer for the ranch, and in spring 1910 Thomson mailed a 60-day option to a Chicago man named Collins asking for $1 million. Collins failed to come up with the money.

By the end of April 1910, Greaves and Thomson were at the end of their tether. Thomson, acting for both, wrote an emphatic letter to Ward.

Sometime in Jan. J.B.G. was here when he stated that he would not go on with the D.L. business under present arrangement, that expenses are too high and he could get a man to do the work @ $40 per month. As he is quite decided to give up it is useless at present to go to the expense of building. He had advanced $4,000 to purchase a Saw Mill to cut Lumber for Ranch purposes and leaves the question to Buildings for future proprietors. He says we never consented to the arrangement with Frank. You took Frank to D.L. & installed him & afterwards reported at Victoria. In short J.B.G.'s proposal is to buy or sell on a basis of $800,000 at which figure he would prefer to sell. He leaves me free to act and adds 'anything you do, I am with you'—but the effect of keeping Frank on is to put him out—it is useless for you and I to try to carry on without J.B.G. . . . I think J.B.G. has been seriously annoyed as . . . there are quite a number of Brokers writing & talking about the sale of the property and he is sick of having people go up to inspect. . . . the unalterable position is that J.B.G. intends to give up this season. At my time of life I would rather go out when Greaves retires & will advocate the terms he proposes. J.B.G. says also that he will not offer his interest to any outsiders until we have fair show to deal with the property. At same time tho' he has quoted his figure. It is understood that any outsiders are not to be allowed to pick up the property at less than the price agreed on. All this is very serious for all of us but it is the result of forcing Frank into the management.

This letter put Ward on the spot and he immediately booked passage from Britain.

The two older shareholders of Douglas Lake drew up an agreement, on 13 July 1910. It stated that "they will buy from or sell to William Curtis Ward at the price of $266,666 for their entire respective holding of shares in the said Company—the said price being fixed upon the basis of $800,000—for the entire capital of the Co."

Ward, with the help of the Canadian Bank of Commerce, decided to buy out his partners. On 3 August 1910, The Douglas Lake Cattle Company, Limited Liability ceased to exist and Douglas Lake Cattle Company Limited Liability took its place with W. C. Ward as its sole shareholder. Frank, who had been managing Douglas Lake alone since June, became manager.

Thomson and Greaves were finally out of the company they had founded. After running the ranch for 25 years, and after being in the cattle business for 43, Greaves had to leave the rolling bunchgrass hills and plains of Douglas Lake, and the way of life he had enjoyed for so long.

Before he left, Greaves presented three of his oldest employees with purses. Joseph Coutlee, who had been riding for him for nearly 20 years, received $1,500; Bob Beairsto, who had been the ranch bookkeeper since 1895, received $500; Billy Fountain, Greaves's favourite Indian pupil, received $500 plus many of Greaves's personal effects, including his saddle, a wagon, some horses, and some farm machinery. It was with great sorrow that all who knew him bade him farewell from the cattle scene. He was 79 years of age.

Greaves moved to Victoria to his new home on the corner of Simcoe and Clarence Streets, with two Chinese servants: one to cook and look after the house and the other to care for the garden. Now and then Greaves returned to the Interior to help Frank Ward and his old neighbours. He loaned Joseph Guichon $40,000 to buy the Triangle Ranch, whose 11,000 acres added so materially to the financial stability of the Guichon Ranch that Joseph repaid the loan within three years.

Slowly Greaves's health failed. On Sunday, 13 June 1915, five days short of his eighty-fourth birthday, J. B. Greaves, ranching pioneer, died. Many were the obituaries praising

Greaves, but it was Frank Ward who summed up the general feeling: "There is no praise too high with which to honour this grand man." When Greaves's will become public, his accumulated wealth rocked the province. He left $605,932: a small amount to his children and well over half a million dollars to charities.

Seven and a half months after Greaves's death, his 88-year-old partner, C.W.R. Thomson, passed away. This was two days after the death of Thomson's wife. They had been married 55 years.

William Curtis Ward grieved to see his business friends of so many years pass on. He was the last survivor of the syndicate and the only remaining founder of Douglas Lake Cattle Company. Yet the seasons on the ranch continued changing, the cattle kept multiplying and fattening for market, more new land became available, and extra fencing needed to be built—and Ward's son Frank was now in charge of Douglas Lake Cattle Company.

The Boar's Nest in winter (courtesy of Brian K. de P. Chance)

Hay camp, 1917 (courtesy of Curtis Ward)

Two A-frame stacks side by side: mass production (courtesy of Curtis Ward)

Driving the bull rake (courtesy of Mrs. Lawrence Graham)

Coming down on slings after stack is finished (courtesy of Mrs. Lawrence Graham)

Grinding grain with steam engine to make chop (courtesy of Micky Lunn)

Resting mower and horses beside Sanctuary Lake (courtesy of Micky Lunn)

Three two-horse teams moving swinging boom stacker to next stack site (courtesy of Mrs. Lawrence Graham).

Threshing machine coming through Home Ranch yard (courtesy of Micky Lunn)

CHAPTER TWELVE

*Our greatest achievement I should say was we lived through many
lean years and kept out of debt.*

Francis B. Ward to Kenneth Coppock, Editor,
*Cattlemen: The Beef Magazine,*13 January 1945

Not knowing the hardships that he and his wife were to face,
Francis B. Ward told his father that he would manage Douglas Lake Cattle Company provided that it was sold at the first
opportunity. Then, and only then, did W.C. Ward sign the
loan at the Canadian Bank of Commerce, and become sole
owner of the ranch on 3 August 1910. His third son was its
second manager.

Thirty-five-year-old Frank was a good-looking man with
clean-cut features. Kenny, elegant and fine of face and
manner, was the first white woman to live on Douglas Lake
Ranch and so had much new ground to break. The couple
moved into the Ranch House while a manager's house was
being built and had barely passed their first winter at Douglas
Lake when their three-year-old daughter, Betty, became ill.
Kenny's parents at Fort Macleod agreed to take their sickly
grandchild until the ranch was sold. This temporary arrangement was to last many years.

In 1912 Frank and Kenny moved into the two-storey frame
house built for them on the edge of the Home Ranch. Green
tiles on the roof, white paint on the outer walls, a brick fire-
place chimney, and concrete steps up to the front and back
doors set apart their large home from the other red-roofed
ranch buildings. An outside well supplied water by means of a
hand pump and a storage tank in the top of the house. The
sawdust-burning furnace in the basement needed frequent
stoking to warm the many spacious, high-ceilinged rooms. A
calcium carbide plant brightened indoor evenings with the
clear light produced by acetylene gas. The house lent itself to

expansion, and later a dining room and billiard room were added. Three dozen Lombardy poplars bounded the garden, and inside space was left for a clay tennis court. A neat white fence surrounded the whole.

Frank mixed the residue from the carbide plant with water, salt, lard and laundry bluing to make whitewash for the ranch buildings and fences. During wet weather or before laying off the hay crew in fall, Ward equipped each man with a ladder, bucket of whitewash and brush, and so kept the ranch immaculately painted.

Paperwork initially absorbed his attention. An up-to-date list of deeded district lots showed the ranch acreage to be over 100,000 acres. As well as the famous | | | horse and cattle brands, new ones were registered—the anchor ⏚ for cattle, and IV ⩒ for horses—so that animals could be marked to indicate that they had been ranch-raised, or bought, and so on.

Frank and his wife travelled around the ranch on horseback. Kenny was a superb horsewoman, stately in her sidesaddle, and her presence put a curb on the cowboys' colourful language and inhibited their relationships with the female native cooks and camp followers. But she was enthusiastic to see range life and to bolster, by her presence, her husband's role of manager.

Frank had the advantage of having worked for the company before, but a lot of new country had been added to the ranch, and there were neighbours to meet as well as new employees. Like Greaves, Frank depended heavily on Joe Coutlee and on Indian labour to get the work done. He now began the huge undertaking of learning the movements of the cattle throughout the year, the times to plow, seed, harvest and feed, and whom to deal with regarding anything from buying a plowshare to selling a team of horses.

The ranch wintered 10,000 head of beef cattle each year, the greater number being commercial cattle ready for market as three-year-olds or older. Two small purebred herds, one of Herefords and one of Shorthorns, made up the total, supplying the commercial herd with some of the new bulls each year. Other bulls came from places as far away as

Ontario. Beef breeding technology was in its infancy; the only known way to put good hindquarters on a Hereford steer was by crossbreeding Hereford females with Shorthorns. Sixty per cent of the bulls used were purebred Herefords; the other 40 per cent, purebred Shorthorns. The cattle thus ranged in colour from the red and white of the Herefords through to the russets, roans and pure whites of the Shorthorns, with many a patchy brockle-face (red freckled) in between.

Starting in summer, grass-fattened cattle left the ranch in drives of 200 that filled ten rail cars at the Nicola station. A crew of eight or more riders took ten such drives—half 1,100-pound cows and half 1,200-pound steers—between June and December each year. Pat Burns's prices remained steady in Frank's first years, at an average of $80 per head, from 6 to 8 cents per pound.

The first two winters impressed upon Frank the importance of having sufficient hay for winter feeding, for consumption still outstripped production. He noticed that the land at Spahomin lay idle yet it was potentially prime hayland, so he encouraged the Indian agent and the Indians to borrow ranch machinery to grow hay. He even loaned them money to buy their own farming equipment. The resultant hay not only fed the Indians' horses and cattle but also provided a surplus that helped Douglas Lake and brought the Indians groceries and clothing from the Douglas Lake store in trade.

The 100,000 acres constituting Douglas Lake's deeded land dovetailed like a vast jigsaw puzzle into the borders of Indian reserve land and neighbouring ranches. Moir and Bond's Norfolk Ranch was to the north. Guichon Ranch and Lauder Ranch were to the west. Earnshaw Ranch and Raspberry Ranch intermingled around Minnie Lake. Innumerable small operations around Aspen Grove shared line fences with more Douglas Lake land. This deeded nucleus supplied fall grazing, winter feed grounds, hayfields and some spring grazing, but most of the ranch's spring and summer ranges belonged to the Crown. Without Aspen Grove, the Lumbum Commonage, Hamilton Commonage, and various private and government leases, Douglas Lake could not have run 10,000 head.

Thus a royal commission investigating the agricultural conditions of the province in 1913 was a new worry for Frank Ward. Homesteaders were pressuring the government to make the provincial grazing commonages available for pre-emption, and it seemed as though the government was willing to acquiesce. The Honourable William R. Ross, minister of lands, considered that the drylands of British Columbia—those areas which received less than 20 inches of precipitation annually—could be more valuable to the province as farms than as range for cattle.

Frank sent a letter to the premier, Sir Richard McBride, and cited the experience of Alberta, whose government had opened grazing land to homesteaders so that the ranchers of the province had had to cut back their herds and many had even been forced to sell out. Now, Alberta was short of beef and the government was offering many inducements to get the ranchers to go back, but it was too late. Three months later, Frank Ward became one of the 33 inaugural members of the Nicola Stock Breeders and Agricultural Association, and he and Lawrence Guichon, as association representatives, tried to persuade the government to farm part of their grazing lands without irrigation and thus prove what they already knew: the land was too dry to farm successfully.

The government accepted the idea and soon fenced in the experimental Quilchena Dry Farm on part of the Hamilton Commonage. Ernie Brookland arrived in the summer of 1913 to run Shropshire sheep there, and to break up the bunchgrass sod to grow wheat, oats, barley and vegetables. He had to contend with the climate in the Nicola Valley: an annual rainfall that locals considered heavy at 15 inches and normal at 10, but which did fall in the growing season; a maximum of three frost-free months a year—June, July and August, and temperatures ranging from a high in June of 93 degrees Fahrenheit to a low in February of minus 26 degrees Fahrenheit.

The experiment would take some years, so cattlemen in the valley settled down to wait for the government decision.

Frank Ward was proving himself a thinking man of sound morals and business ethics. His ability to make a decision and see that his men carried it out correctly quickly gained him respect. His disapproval of any unsavoury or doubtful action by others was immediate, and was usually accompanied by a muttered "tut-tut-tut-tut" and a prompt attempt to correct the situation.

His position at Douglas Lake was still temporary, though, and early in 1914 Frank went to England to see if a sale was likely. He found that although his father had been dealing with many potential buyers, including a Vancouver businessman and a syndicate of London financiers, there did not seem much chance of a quick sale. Part of William C. Ward's initial eagerness to sell was because his purchase of the ranch had put him heavily in debt. But by 1914, ranch profits had paid off over $200,000 of his debt and his need to sell was lessened. Even a cash offer of $1,150,000 was rejected as insufficient.

He now intended to keep the ranch as a nest egg for his large family and discussed the idea at great length with Frank to ensure that both he and Kenny would be prepared to stay on. He wrote to his son-in-law, Will Oliver, in Victoria who was a director in the company:

My idea now is to keep the property as an Investment for the family. Including myself, there are ten of us to share in it, which would give a tenth to each. I should take a liberal salary as Managering [sic] Director. Frank as Manager would get good pay for his work, say $5000 & free quarters .... [He] would have a free hand in control of the management & the certificates of shares would be endorsed & lodged at the Bank in my name. If there was a yearly dividend each would get the dividend available & of course, so long as my debt is not paid off, dividends would not be forth coming, but I look forward to this year's profits to clear that off. One principal advantage would be to save death duties in England, & I should almost dispense with a will.

The profits of just over $100,000 for the year ending June 1914 were insufficient to clear W.C. Ward's loan, and as the company owed him over $800,000, investment and interest, it was incredible that he expected them to do so. On 25 June 1914, he signed an indenture releasing the company from this debt in consideration of its undertaking to pay him an annual sum of $25,000 for life. The same day, redistribution of the 400 shares began and Douglas Lake Cattle Company became a family concern.

His decision to hold on to the ranch was a good one, for the long-expected war against Germany broke out, causing cattle prices to soar from $80 a head in 1914 to $111.43 in 1919. Annual dividends from 1914 to 1920 averaged $68,000, and that was after money had been set aside for the purchase of new land. This proved to be the most fruitful period of the Ward family ownership.

Yet even as these profits were being made, Frank Ward's problems were increasing. The half million Canadian men who had gone into the army made labour scarce. Many a big ranch resembled an old men's home as it struggled to operate with only a skeleton crew of men too old or too unfit for military service. Competition for the smaller labour force meant that higher wages had to be paid. Beaver populations increased because they were no longer being trapped, wreaking havoc on Home Ranch, Chapperon and Minnie Lake hay meadows. For these and other reasons, ranch operating costs began to increase at the staggering rate of $15,000 a year. In 1914 operating costs were less than $70,000; in 1920, $160,000 barely covered them.

Some Douglas Lake cattle went to war-torn France in 1915, when Patrick Burns bought a trainload of fat cattle from Nicola Valley ranchers and shipped them live across the Atlantic, the boat dodging submarines and warships. It was a major gamble and Pat Burns spent many an uneasy day until the cattle arrived safely. Never again did he go so far west to fill his war contracts.

The year 1917 brought fresh worries for Frank Ward. The Canadian government brought in a temporary wartime measure—the Income Tax Act. The Spahomin Indians had a disagreement with their Indian agent and asked Frank to intercede. A new Brand Act was under consideration and Frank had to devote time to discussing it. Coyote bait put out by the Indians poisoned his puppy, Shelly. The Douglas Lake road, a public artery, was in such bad shape that the bridges and culverts could take only half a truckload, and Frank told the Merritt road superintendent that "unless something is done very shortly there will be a nasty accident some night in the dark."

Victoria set aside no money for such roadwork for many years, however, and many were the times that Frank's Studebaker would get stuck in one of the mud holes, forcing the ranch manager and his passengers to spend all night inside the car or at the Morton cabin. In the morning, Frank would take from his car a portable cable that he could throw over the government-owned telephone wire strung along the roadside fence; this connection enabled him to summon a team of horses from the bookkeeper in the store office to pull him out.

Douglas Lake lost a sizable portion of summer grazing in 1917. The previous year the ranch had paid $200 for depasturing (grazing) 2,000 head of cattle on the Lumbum Commonage. Other users of this 30-year-old 25-square-mile tract of land lost similarly, for Grazing Commissioner Thomas MacKenzie retitled it a grazing allotment and granted it exclusively to Jack Whiteford, who had recently bought a ranch neighbouring the Lumbum. Ward reworked the summer grazing pattern of Douglas Lake stock to adjust to this change, and worried lest the experiments at Quilchena Dry Farm prove successful, for then the Hamilton Commonage might be the next to disappear.

Also in 1917, the Kettle Valley Railway Company expected the stockmen sending shipments from Nicola to water their animals in a creek in the centre of the busy village, which was unnerving for cattle. To avoid a stampede, the stockmen loaded their cattle without watering them, causing their stomach and tissues to shrink. P. Burns and Company complained sorely about this weight loss, even offering an extra dollar per head if Douglas Lake would go back to driving its cattle the longer distance to the Kamloops station.

With labour so hard to get, Frank had to answer a job inquiry, "With reference to your friend Walker [of the Chartered Bank of India, Australia & China], we should be only too pleased to have him here, but we have no accommodation but a bunk house which consists of every sort of people who always carry their own bedding, and our cow camp is the toughest in the country being mostly Indians and breeds . . . . Of course if he thinks he can face a bunk house, he is welcome to come, but I would hate to tackle it myself."

For Kenny Ward, life was easier in 1917, when Frank hired Nellie Guitteriez, a Spahomin girl. Living in the laundry, a small frame building close to the Wards', Nellie took over much of the housework, allowing Kenny to devote more time to her husband and visitors. The Wards entertained extensively each fall, for during the duck-hunting season, Frank left the day-to-day running of the ranch to his foremen and turned sportsman. Kenny turned chatelaine and was busy preparing picnic lunches and roast dinners.

There were six shooting areas plus the Upper Nicola reserves, which were included by permission of the Indians. Thus each day for a week the Wards and their guests could shoot fresh lakes and by the following week, when the next visitors arrived, the birds would have settled down to provide more good shooting. From mid-September to the end of October, when the birds flew south and the lakes froze over, hunting took precedence.

Once, with his house full of hunting guests, Frank decided to shorten his thorough nightly inspection of the almost-empty horse barns. It was a fall evening after the war; the cowboys were at camp and the haying teams were still out. That night, 20 tons of well-cured hay ignited with a roar inside the polo barn at the Home Ranch. The only horses there were the Wards' personal mounts, but those three cherished saddle horses died in the blaze. Kenny grieved their loss long after the new barn was built and in use. Ward attributed the cause of the fire to a smoking drunk sleeping it off in the hay, but he knew that the danger of fire would always exist while the men carried coal oil lamps to attend to their horses.

Myriads of grasshoppers caused chaos with the bunchgrass ranges and hayfields of the Nicola Valley during the war years. The ranchers were unable to combat the plague by the proven method of plowing up the grasshopper breeding grounds, for they were on the delicate bunchgrass ranges. The country looked more like an African desert each year. In order to acquire extra grazing, Douglas Lake bought Archie, Fred and Patrick Raspberry's 8,000 acres at Minnie and Pennask lakes.

By 1918, two years later, it became quite clear that to avoid permanent damage to the range from overgrazing, the herd size would have to decrease. Fortunately, prices were rising and good returns were obtained from the extra 750 animals going to market that fall.

To maintain a constant herd size, a rancher must sell the same number of fat cattle each year as the number of calves that are born. To get Douglas Lake's herd size back up to 10,000, Frank had to sell fewer cattle than the number of calves born, and of course the calf crop was already even smaller than normal because he had sold some of the producing cows. But even such a slow method of recovery was impossible, for the numbers of grasshoppers did not decrease. The ranges continued to deteriorate, and the herd continued to diminish through further culling.

War, labour shortage, grasshopper plagues, reduced grazing—then the influenza epidemic, which would eventually kill 20 million people throughout the world. It struck the Interior of British Columbia in the fall of 1918. Kamloops was ready, for the chairman of its Hospital Board had seen the sickness rampant in eastern Canada. Hotels, churches and vacant rooms became emergency hospitals in Kamloops and Merritt. There was little resistance to the 'flu, and whites and Indians alike fell like flies. Kenny Ward helped where she could, tending the sick and taking hot food to the Indians to increase their resistance. But the fatalities continued and many a farm labourer put in a full day burying victims.

As the epidemic reached its height in the Interior, the long-awaited news came that Germany had signed an armistice. On 11 November 1918, the war ended. Mrs. William Munro, postmistress at Nicola, gave the familiar five telephone rings that brought everyone to their phones to hear of the end of the war. Celebrations were held throughout the Interior, and, despite the year-old Prohibition Act that was ruining such hotels as Guichon's Quilchena Hotel, many a secreted bottle of hooch surfaced for the occasion.

Gradually the war troops came home. Their return should

have ended the labour shortage, but when Canada rewarded her soldiers with a free land grant of 160 acres, many chose to settle and work their own land. As Frank Ward had said seven months before Armistice Day, "the only labour we are able to get here are the Indians, but they are now much in demand, and are getting very independent."

Frank Ward played an active role in uniting ranchers in the province. More and more, cattlemen were realizing they could not fight their battles alone against such groups as the government, railway companies, meat packers and consumers. Many of the associations that sprang up across British Columbia had similar names and purposes and succeeded mainly in confusing their results, yet on the whole much good came from them.

Frank helped the Cariboo cattlemen start their own association in 1914. The same year he became a director of the newly formed B.C. Stock Growers' Association, and a delegate to the Interior Stock Raisers Association of B.C. which joined with the B.C. Stock Breeders' Association in 1919. In this capacity he became a B.C. representative to the Western Canada Live Stock Union, which had its head office in Calgary. In 1920 he became the founding president of the B.C. Hereford Breeders' Association.

The provincial Department of Agriculture joined with the B.C. Stock Breeders' Association in 1919 to hold the First Annual Bull Sale. Held in Kamloops in early spring—the time when western Canadian ranchers are sizing up their bull herds to decide how many new bulls are needed—the event included judging purebred Hereford, Shorthorn, Angus and Red Poll bulls and fat stock and then auctioning them off. Frank Ward was a director of this First Bull Sale and supported it by bringing along 30 of the ranch's purebred bulls. The event was an unprecedented success, people coming from near and far to buy and sell.

For Douglas Lake, it was not a complete success, for Allan Baker of Loon Lake, Clinton, wrote to complain that one of the Douglas Lake bulls he had bought had developed only one testicle. This exemplified how unsophisticated ranchers then were in assessing animals at auction. Frank's reply was honest and charming.

126

These bulls had a very hard deal before going to the sale, and were picked out in very cold weather, and driven over to Kamloops, a distance of 75 miles through deep snow, and as the cold has, as a rule, a decided effect on the testicle of a bull, it is not so much to be wondered at as you think, that such a thing might be overlooked. If the bull is wrong, we will appoint somebody to examine him, and will replace the bull by exchange next spring or refund your money and turn the one you have over to the Butcher. We regret very much that such a thing should have occurred.

His straightforward approach humoured Baker. "Your courteous letter has completely disarmed me . . . . Never mind about the bull; let it go. I have turned him out on the range & have no means of knowing if he gets calves or not. I have seen him cover a cow in an unorthodox fashion. I deserve all I get for my carelessness."

In 1919 the first airplane to land in Merritt drew big crowds. Pilots Hoy and Dickson were organizing a Vancouver-to-Calgary mail route and were acquainting themselves with the people and terrain over which they were to fly. Frank Ward flew over the ranch with them, this being his first time in a plane, and spotted a band of Indian horses enjoying company grass.

Although Douglas Lake Cattle Company's 121,000 acres were off the market, managing director W.C. Ward was still receiving sale offers. In 1920, one came from the States for $1.5 million. The herd was still declining, down to 8,200. Cattle prices were good, though they had dropped $6 from their peak in 1919. The horse herd was up to 900, partly because the horse market had slowed down. The grasshoppers had been on the ebb for some time and the ranges were improving. The 36,000-acre Hastings Ranch south of Merritt had increased the grazing leases that year. To the senior Ward, $1.5 million was not a good enough price, and he asked for $1.75 million. It was more than the buyer could handle, and yet another deal was off.

Rather than sell Douglas Lake in 1920, W.C. Ward approved the purchase early in 1921 of Moir and Bond's 5,000-acre Norfolk Ranch. With this land, which included some much needed haylands, came 80 horses, a herd of cattle

and various buildings. Talbot Bond stayed on to run this area
north of Chapperon. The purchase price was $75,000, Moir
and Bond receiving $40,000 and Douglas Lake assuming two
outstanding mortgages for the balance.

Douglas Lake reacquired land in 1921—the Quilchena Dry
Farm. For a total of eight years, experiments had continued
on the Hamilton Commonage with varying degrees of
success. Then in 1920, just as a good crop of oats was
maturing, a plague of grasshoppers consumed the entire
year's work. The provincial government promptly abandoned
the Dry Farm with its plowed fields, fences, living quarters,
sheep corrals and sheds. Coutlee's crew benefitted by using
these simple buildings during turnout each summer thereafter. Ward and his neighbours heaved a sigh of relief to see
the Hamilton Commonage remain a valuable grazing range.

Even so, they wondered if the farm would have succeeded if
there had been no grasshopper attack. They did not have to
wait long to find out. The Land Ministry threw open to homesteaders various grazing ranges in the province, including the
Knutsford plains area south of Kamloops. Each preemption
there covered only 160 acres. The Nicola Valley ranchers
watched these homesteaders lose all their savings as they tried
to eke out a living from the dry soil, and knew that the Dry
Farm experiment had been confirmed.

As shrewd a businessman as he was, not even W.C. Ward
could have foreseen the depression that began in the summer
of 1921. If he had, he would have sold Douglas Lake in 1920.
Overnight, or so it seemed, cattle prices fell from $105.45 in
1920 to $58.65 in 1921, the grasshoppers returned in full
force, labour remained difficult to attract and keep, and the
herd size continued to shrink.

In the depths of this crisis, the managing director of the
company became extremely ill. Since moving back to Victoria
from England in 1915, William Curtis Ward had received
several blows: not long after he had contracted a serious illness, both his wife and Will Oliver had died. W.C. Ward went
to Paso Robles, California, to improve his poor health. Instead, he died there on 12 February 1922, aged 80.

Even though he had transferred nine tenths of his Douglas

Lake Cattle Company holdings to his children, W.C. Ward's estate in England and British Columbia was still worth almost $200,000. If the 1914 redistribution of shares had not taken place, he would have died a paper millionaire, yet without the funds to pay the probate and succession duties.

The death of W.C. Ward caused Hample, the American who had offered $1.5 million for the ranch two years earlier, to reopen negotiations. Despite the addition of the Norfolk Ranch, he now valued Douglas Lake at a little under $1.3 million because of the drop in cattle prices. Frank Ward was nearing 50 and beginning to feel very tied to the ranch. The month and a half of duck hunting each fall was his annual holiday, for he found it difficult to leave the ranch for business or pleasure, as there was no one capable of running the place in his absence. Polo was being played again after it had stopped during the war, and for this and other reasons Ward wanted more freedom for himself and his wife.

Since Will Oliver's death, Jim Abbott, a Vancouver lawyer and old school friend, was the new cotrustee with Frank under the 1914 settlement. The two analyzed the worth of the ranch and, though their valuations differed in make-up, they agreed that it would be wise to sell for a figure somewhere between $1.3 million and $1.5 million.

The Trust Deed gave the trustees and executors of W. C. Ward's estate absolute authority to dispose of Douglas Lake as they saw fit, and to reinvest the money in more stable securities to gain a fixed income for the beneficiaries. Yet Jim Abbott, Frank Ward and his brother, Willie Ward, an executor, wrote to the rest of the family in England to hear what they thought of a sale at the price range agreed on.

All five met in Sidmouth, England, to deal with the matter. Led by their brother, Cecil W. Ward, an eternal optimist and still something of a promoter, they wired back, "Family do not approve of sale for less than one million five hundred thousand dollars net cash free from all commission." They agreed it would be a great mistake to sell at a sacrifice—at a figure which they knew had not been acceptable to their father two years earlier. Their advice to Frank was to find himself an assistant, and to take life more easily until things improved. "We can then get our figure .... Until the Ranche is sold we would be most grateful if the Board would when possible put funds to reserve so as to maintain regular dividends, none of us have *sufficient independent means* to live on *without using* what we get from the Company."

In the summer of 1923, Frank and Kenny travelled to England to see their daughter Betty, who was going to school there, to have a holiday and to try to sell the ranch. H. Hamilton Abbott, who had some ranching experience, managed Douglas Lake in their absence, for as Jim's brother he had Ward's trust. It was a good holiday, but no buyer was found who would put up $1.5 million.

Bustling 56-year-old Major Charles S. Goldman appeared in 1924 as another purchaser. He held an option on the ranch for something over a million dollars, an option which he did not take up, but he did buy other properties in the Nicola Valley. A South African, the major was an influential man, later becoming secretary-treasurer of the British Empire Emigration Department and a member of Parliament for Penrhyn and Falmouth. He owned quite an empire, including diamond mines, a Swiss hotel, and sisal and peanut plantations in Africa. His wife, Agnes, a Dame of Grace of the order of St. John of Jerusalem, was the daughter of a British baronet and a descendant of Sir Robert Peel. The major, his wife and sons drove their Cadillac into Nicola in 1919 to take possession of the Lakeview Ranch, a property at the south end of Nicola Lake which they had acquired during the war.

Major Goldman quickly added more properties: Voght Valley, some Nicola ranches owned by the Marriots, Winneys and Clarkes, some of Douglas Lake's less advantageous lots near Nicola, and even the townsite of Nicola from the CPR. His initial plan was to divide the land into small holdings and to bring out a number of English families to live on the previously surveyed townsite of Nicola. This type of plan had failed in the valley once already when the English syndicate that had purchased the ranches of John Moore and Robert Scott went into receivership in 1915. Acting on behalf of himself and his six brothers and sisters, who now owned the large ranch built up by his father, Joseph Guichon, Lawrence Guichon in 1920 acquired Moore's 20,000 acres—the Beaver Ranch. Goldman's plans also failed, but instead of giving in, he incorporated Nicola Lake Stock Farms Limited, which became one of the bigger ranches in the valley.

Each year Douglas Lake stock summered with Guichon stock west of the road between Merritt and Princeton. The cowboys drove the cattle through Kane Valley and scattered them around Voght Valley, Kingsvale and Brookmere. However, soon after Major Goldman had purchased the deeded land at Voght Valley, he applied for the surrounding 28,000 acres of summer range already held by Douglas Lake and Guichons, and received it. Ward and Guichon lost summer grazing for around 800 head. Douglas Lake continued to share the grazing south of Voght Valley with Guichon Cattle Company, and Ward began to improve the forage closer to the Home Ranch, up Mellin and Beak creeks.

With the town of Nicola in the hands of one man, the population decreased, and the CPR served notice in 1923 that they were going to close the station. Major Goldman organized public meetings to protest this move as being detrimental to the ranching community. The station remained open, for a year. In 1924, the CPR again served notice; passenger service from Nicola stopped, and the station agent, Reid Johnston, moved elsewhere, but, thanks to Major Goldman's tenacity, freight service continued for those willing to do their own loading of cattle cars. Every rancher in the valley loaded his stock personally from then on. Despite its many setbacks, the town of Nicola, though finally a company town, remained the agricultural hub of the valley.

The Douglas Lake basic herd—the herd base which produced the annual calf crop—reached its all-time low in 1922 at 7,500. Frank Ward began the monumental task of increasing the herd. The ranges were poor and would not support a sudden influx of more cattle. Each heifer that was held over to enlarge the cow herd tied up $60 that could have been obtained if the heifer had been sold. No dividends to the shareholders could be paid.

Systematic cattle rustling added to the difficulties. The

cattle counts at Douglas Lake for 1923-24 and 1924-25 showed a total shortage of some 1,200 head. Willie Ward, the Vancouver consul for Denmark and Iceland, solicited help from the U.S. Consul General. A convicted cattle thief, a companion and "a man called Jim Turner . . . are supposed to have moved 600 head through by Savonas to Mamette Lake and there to the Aspen Grove country, and so on, down into the U.S. side, along with our cattle and Guichons," Willie wrote in August 1925. Jim Turner, a one-time Minnie Lake rancher and the husband of Minnie Earnshaw, had been foreman for C.E. Wynn Johnson of Alkali Lake Ranch. Then he left for the States and "had about $3,000 or more in his possession when he left, being the proceeds of the sale of some of these cattle." Willie continued, "My brother says that in addition he had money belonging to settlers in The Chilcoten who gave him money to buy cattle for them and they are now looking for him . . . . From what I understand, there is this secret trail . . . through the mountains . . . . The corrals have been found where they put these cattle for the night, both on your side and the B.C. side . . . . We know that during the war quite a few horses of ours were stolen and taken over."

Moccasin telegraph had found Jim Turner guilty, but he never stood trial for the offences. The cattle rustling dropped considerably.

Apart from keeping extra heifers back to rebuild the herd numbers, Ward bought, in 1924, 377 head belonging to the Land and Agricultural Company in Vernon. With them came their brands, $\sqsubset$A and $\rightthreetimes$. In early November 1925 he bought 259 head of cattle from William Campbell of Cherry Creek. He also bought Campbell's brands: $\cup$+ and $\curlyvee$+ on left ribs. The brand $\boxminus$ came to him through the clearance of another herd the following year. Having written to Calgary in 1925 for 200 calves and finding the price there too high, Frank sought yearlings the next year and was no more successful, although he did get 165 calves through Pat Burns.

Also in 1926, he bought 17 bulls from W.W. Sharpe of Stettler, Alberta. Frank bought the animals blind and wrote, "I would have much preferred to have seen what I was buying, I am simply going on what I already know of your herd. I do not like white bulls, for more than one reason, and that is with a white sire in a mixed herd you are much more likely to get off colours than from roans or reds. Also the range man has christened them the Green Durhams, for if there is a lousey one it generally is white. That may be simply that he shows up on account of his colour. I notice that four of yours are white?" Despite his misgivings, he was actually delighted with the bulls once he received them, for they fared very well.

That same year, Billy Lauder offered his entire herd—690 head—to Douglas Lake at $30 each. Ward wrote to the company's bank manager and his codirector Grange V. Holt, "It looks as though the Guichons and ourselves will be the only people with many cattle in this district before long if this man sells." He took Holt's advice in this matter and decided not to buy right away but to see how the rest of the year went first, a decision that kept Lauder in the cattle business.

The bottom fell out of the horse market in the '20s. Since the purchase of The Boss in 1888, Douglas Lake had sold Clydes as teams for ranchers, farmers, loggers and milkmen, but everything was becoming more mechanized and the demand for heavy horses had dropped. The Clyde registration book lay idle, and the herd diminished little by little from a high of 900 to half that figure. The vacated horse ranges became cattle ranges.

Ward did not have an assistant manager until 1926. Brian Chance came from Australia as a 21-year-old with experience working on sheep and cattle stations. It was not his first time at Douglas Lake, for he had worked there briefly in 1921, but even so he had a great deal to learn. He first went to work with Joe Coutlee.

In 1926 the grasshoppers returned and spring droughts bared the range until it resembled a wilderness. The remaining 1,200 head of summer beef were grazing the last feed available by July. Desperate for an outlet, Ward urged Blake Wilson, Pat Burns's Vancouver buyer, to take more cattle, but Wilson had his own headaches and took even fewer than the first figure discussed, which further upset Ward's plans.

Also, although the hay crop was ready for stacking, every

man was out fighting the many forest fires burning throughout the valley. At least 300 men were fighting fires up the Salmon River above Norfolk, on Quilchena Creek, and at Kingsvale in Aspen Grove. Some Douglas Lake cattle burned to death in the fires.

As the 1927 New Year got closer, it became obvious that there was not enough hay for the winter. In addition to distributing cattle between Norfolk, Chapperon, Minnie Lake and the Home Ranch, sending a few hundred head to Westwold, and buying all the available hay at Spahomin Reserve, the manager had sent cattle to Lower Nicola, over 40 miles away, to eat all the hay the company could buy there.

As the cowboys drove these cattle to Lower Nicola, the trustees for Douglas Lake and the executors for the Ward estate renewed their discussions about selling the ranch. "Under existing conditions I find it hard to arrive at a true value of what we have," wrote Frank to his brother, Willie. As a going concern he thought it worth between $800,000 and $900,000. Willie valued it at $750,000, and would have been happy to take his share if that figure were only $600,000, saying, "The truth is the place is too large. . . . The buyers who have the cash to take over this place are few and far between, and as you know, can do better with their money."

The trustees considered many options: selling the ranch as a going concern; breaking it up and selling it in parcels; selling the land and cattle separately. Willie Ward asked Frank about the value of the ranch timber, because Samuel Ryder, owner of the Hastings Ranch, which Douglas Lake leased, was selling the timber on his 36,000 acres for $800,000. Ranchers were even rumouring that Ryder was going to make a bid for Douglas Lake with his timber money. But when Greaves and Beak put Douglas Lake together, they had bought as much timberless land as possible, and Frank could only guess wildly that there was somewhere between a half million and 10 million board feet of coniferous timber growing on the deeded land, valued at between $1,000 and $20,000.

Finally the trustees agreed to sell the ranch for $800,000. Despite his own advice to wait until summer, Frank immediately offered it at that price to Pat Burns. This figure was $10,000 less than the sum Burns had offered Greaves, Thomson and Ward 18 years earlier. The 72-year-old packing house and ranching tycoon replied, "If I were forty or fifty years younger there is nothing I would like better, but I am afraid under the circumstances I would not be interested."

In May, Frank found a buyer, but the buyer's motives did not impress Grange Holt. "The idea of even a New Yorker, wealthy as he may be, considering buying a ranch like Douglas Lake at a cost of $750,000 for fishing and shooting purposes, seems to be absolutely preposterous.... if the approximate price you quote does not scare off the nimrod, why I suppose we shall have to give it further consideration, but certainly it will be the dearest shooting I have ever heard of."

Douglas Lake was not the only ranch for sale in 1927. Aspen Grove Land Company, set up in 1912, comprised many thousands of scattered acres. Peat, Marwick, Mitchell & Co., acting for the absentee Scottish owners, offered the Aspen Grove Land Company to Douglas Lake, but the bigger ranch was unable to take advantage of the offer. Instead, they began leasing the land for its summer grazing.

The turning point of the poor price cycle came in 1928, and, according to Frank Ward, "the range never looked better since I came here." The purchase price rose to $1 million once more as cattle and land values increased. More and more potential purchasers inspected the property. The trustrees even approached Pat Burns again, this time showing him several balance sheets, but Burns did not make an offer.

One bitter day in February 1929, some of the Home Ranch crew discovered that Bob Beairsto had died a day or so before in his cabin, his home of 34 years. When Ward inspected the shack, he discovered a trap door that, when raised, revealed a cellarful of empty liquor bottles—the accumulation of years. Ward was shocked to realize that the quiet, seemingly abstemious Bob had been secretly drinking on the premises for so long, against his orders—orders which Ward himself did not follow. Some weeks later, Brian Chance, Ward's assistant manager, moved out of his room in the Ranch House and made Bob's one-room shack his Home Ranch base.

Despite the eagerness of the trustees to sell, the ranch was constantly being improved. Bad mud holes, where in dry years animals had been trapped and starved to death, were fenced off. Arsenic bait seemed to have temporarily solved the problem of the grasshoppers. Financial aid from the government helped open up more cattle trails so the cattle could reach the feed within Douglas Lake's heavily timbered grazing permit range. The British Columbia Beef Cattle Growers' Association brought all the smaller associations under one umbrella in 1929 and Frank was elected its first president—a position he was to hold for 11 years. Douglas Lake changed from a private company to a family corporation in order to save Dominion Income Tax.

Frank and his cotrustee, Hon. Fred J. Fulton, K.C., who had replaced Jim Abbott after his death in 1927, received a letter from Cecil Ward and the other English shareholders dated 21 October 1929, advising against selling Douglas Lake for less than $1,150,000 in view of the rising cattle prices. But this was followed by letters from two of the sisters saying that they would accept less than $1 million, if they could have a fixed income. Cecil was in the middle of further negotiations to sell Douglas Lake in England when Wall Street crashed on 24 October 1929, and he abandoned his efforts.

Though Frank Ward managed the ranch and was one of the trustees, the shareholders were breathing down his neck at every move. In a letter of 10 December 1929 addressed to "My dear Brother and Sisters," he pointed out the difficulty they placed him in.

We have had many practical, wealthy Stock men go thoroughly over the property, men such as Pat Burns who recently bought all the big Ranches in Alberta, including the Bar U. He says that nobody could make interest on anything over $800,000.00 and that the Douglas Lake Ranch is not worth more. Mahoney Bros. of Spokane and Montana who have been ranching for forty years expressed their opinion. It was the finest Ranch they ever saw, but could not pay a Million and we would not take less.

We were negotiating with some Americans and had given them to understand an offer of a Million cash would buy the ranch. In view of your letter, we wrote to Clark.... I fear this may break off negotiations.

My reason for wishing to sell: Civilization is gradually killing us. It will never help us, as is the history of all large Ranches. You must realize I will be the loser, we lose our home and salary.

I fail to understand any of you; you evidently have confidence in me to carry on, but none when I distinctly tell you we should sell at a Million if we can get it. Did it occur to any of you to put yourself in the place of a purchaser? With a million to invest, would you choose... a Cattle Ranch, when any good Municipal Bonds will give you the same return, with none of the risk?

When the American stock market disaster occurred, Douglas Lake was getting the best prices in nine years—$102 per head. The ranges were so much better that the ranch sold 880 fewer cows than usual, for Frank was still struggling to get the herd numbers back to 10,000. In spite of the smaller returns for the year, the trustees declared an interim dividend of $8,000 in December 1929, as Cecil and Willie Ward were very "pressed for money."

Douglas Lake entered the '30s with a herd of 8,500 and saw the cattle price dropping from $102 in 1929 to $37 in 1934, $21 lower than the lowest price in the '20s. Though P. Burns and Company were still taking 70 per cent of Douglas Lake's salable cattle, Swift Canadian Company started taking up to 25 per cent, while local butchers took 5 per cent. Despite the low prices, the herd increased, helped by the demand for lighter cattle so that two-year-old steers could be sold at 1,000 pounds rather than three-year-old steers at 1,200. It had taken 5 years to deplete the herd by a quarter; it took the next 11 to build it back up. In 1933 the herd size reached 10,213 and stayed there, an incredible feat, totally attributable to Frank Ward's sagacity and ability. The cries from the shareholders increased in direct proportion, though, for times were tough and dividends inadequate or nonexistent.

As well, the ranch received many setbacks and met many extra expenses from income during the tough '30s. Sportsmen and campers often ignored the ranch's "No Trespassing" and "No Shooting" signs to hunt, chasing cattle away from their watering holes. Ward complained to the *Kamloops Sentinel,* "We found cattle shot, killed, others maimed, horses driven through wire fences by fright from either cars or shooting.

... Do they consider we (stockmen) should sacrifice our business for their pleasure?" Cattle rustling returned on a small scale. One former company cowboy felt so guilty after slaughtering a Douglas Lake steer for beef that he shot himself in his cabin near Aspen Grove. That area became known as Suicide Valley.

The original Ranch House caught fire one cold and snowy December day in 1931 when the stovepipes leading up from the dining room overheated and the surrounding wood smouldered into flames. A new Ranch House, almost identical to the original one, was built and later called the cookhouse. Forest fires in 1934 burned more than a dozen of the ranch's 300 miles of fence and each rebuilt mile cost $250.

For three summers, Douglas Lake regained use of its old grazing permit at Voght Valley because Major Goldman was unable to keep it stocked. They lost it back to him in 1934, straining neighbourly relations as his herd had not increased.

When in 1935 Brian Chance married an Australian girl, Audrey, a two-storey, five-bedroom, gabled house was built for the couple opposite Sanctuary Lake and the Home Ranch yard. Frank moved his office into the old shack where Brian had been living.

The 1936 purchase from H. Hamilton Abbott of the 1,000-acre Earnshaw Ranch at Minnie Lake was one of several that increased Douglas Lake's deeded acreage and, equally importantly, its land taxes. A further 1,600 scattered high-country acres were bought from Wayne Sellers. Diamond Vale Coal and Iron Mines, which owned 4,000 acres on Quilchena Creek, allowed their tax payments to lapse, and Douglas Lake bought this acreage at the 1938 tax sale.

To survive the tough times, Frank Ward cut expenses in every direction. Directors' fees went down 50 per cent to $125 annually. Two foremen who employed teachers at home for their children—Lawrence Graham at Chapperon and Talbot Bond at Norfolk—received notice that the ranch could no longer pay the teachers' wages, nor their board. Nellie Guitteriez, Kenny's home help, was paid from Frank Ward's personal account rather than from the company's.

Food items remained simple: beef, bacon, red and white

beans, potatoes, vegetables in season, dried fruit, milk, dairy butter, tea, coffee, salt, pepper, flour, sugar, baking powder, yeast, lard, Rogers Golden Syrup, and rice for the Chinese employees.

Brian Chance pooled funds one day with the ranch accountant, Ernest W. Chamberlain, and the Home Ranch foreman, George Leith, to buy a case of eggs because they had not had any in such a long time. Frank Ward looked through the store a while later, and left a note for Chamberlain. "I see in this cellar—case [of] eggs—who instructed purchasing these? I shall insist if these practices continue to read every order & to see every traveller. . . . Plain simple food was always our motto. We shd provide all necessity off the ranch or go without."

Home Ranch, Chapperon, Norfolk and Minnie Lake each had its own dairy herd, which provided milk for the orphan calves, and milk and butter for the crews, and at Chapperon, Lawrence Graham raised pigs for pork and bacon; in addition, each area grew a vegetable garden. But Frank Ward would not permit laying hens, so poultry and eggs were scarce at Douglas Lake. Ward saw nothing incongruous in denying his men eggs though he was sharing with Guichon the cost of buying a dozen eggs a day to feed the stallions in the government-sponsored co-operative horse breeding camp at Stump Lake.

To reduce Douglas Lake's expenses, Frank Ward successfully persuaded the superintendent of lands in Victoria to reduce the fee for grazing leases from 8½ cents per head per acre to 4 cents.

Once offered hay at $6 a ton in 1934, Ward replied, "We are not buying hay unless we can buy for just half your price. The price of cattle will not allow us to pay more. It is horrible." The horrible prices then were $2.50 per 100 pounds for steers, $1.50 per 100 pounds for heifers and 75 cents per 100 pounds for cows.

The demand for Douglas Lake's excess purebred Shorthorn and Hereford bulls—unpapered because Greaves had never bothered to register the purebred herds—picked up in the '30s. Ward sold around 30 head each year: calves for $25, yearlings for $50, coming two-year-olds (18 months) for $75, two-year-olds for $100.

Frank avoided paying cash whenever he could, preferring to trade. In 1933 he exchanged three ranch bulls for three belonging to a rancher in Squilax. That same year, a Lytton rancher wrote, "I have got lots of Alfalfa seed No 1, and 2 or 300 boxes of Golden Russet apples which I would like to trade for a bull 2 or 3 years old and a heavy young work horse trained to harness."

The trading of goods did not stop at apples and bulls. In May of the same year, a Vancouver realtor unsuccessfully tried to exchange Douglas Lake Ranch for a nine-storey building in St. Paul, Minnesota. Potential buyers continued to appear, but their offers dropped considerably in the '30s, to as low as $300,000.

In October 1936, the trustees of the Ward settlement and the directors of Douglas Lake Cattle Company met to discuss the future of the ranch, for it seemed gloomy. Government grazing was becoming more expensive and harder to obtain. Taxation was heavy and threatening to become more so. The danger of infectious bovine diseases was constant: brucellosis, a bacterial infection that causes abortion, kept the calf crop at 70 per cent; coccydiosis and calf diptheria threatened live calves; tuberculosis took its toll of mature stock. The next grasshopper plague might be resistant to arsenic bait. Forest fires were a continual worry.

P.R. Duncan, secretary, sent out a detailed document to all the shareholders in November, encouraging them to agree to a sale of Douglas Lake Cattle Company for $600,000 net if a buyer came forward. This could provide an annual income of $2,400 to each shareholder, compared to the $1,000 average annual return over the previous 15 years. Cecil Ward wrote back, "To sell at such a sacrifice: viz: for a *knock out* sum of $600,000[00] nett would be too disastrous for words and especially when things are beginning to improve. We feel that a very strenuous effort should be made to reduce the overhead expenses and thus try and put the concern on a paying basis again. After which it might be possible in a few years time to sell at a reasonable sum."

Each time the trustees and directors had agreed upon a price, Cecil Ward had claimed that the price was too low and declared that Frank was being disloyal in sacrificing "Father's life work." The truth was that W.C. Ward's involvement had been confined to founding the company and buying out his partners. He had always worked for the Bank of British Columbia, which rewarded him well for his excellent service. Never during his 38-year association with Douglas Lake had he worked at the ranch nor struggled with adversity and price drops, nor had he been forced to account to the shareholders when no dividend was forthcoming. Douglas Lake Ranch was much more Frank Ward's life work than it had been his father's.

Cecil's idea was to fire all the "expensive foremen" employed at the ranch, but even he realized this would be hard on his brother, who was so friendly with them.

The difficulty could be overcome by your retiring from the management and appointing Brian Chance . . . in your stead . . . . [He] could then make the changes . . . and being a young man could stand the strain of the very hard work that this campaign would entail and by giving him, say, $240⁰⁰ fixed salary per annum and a bonus of 5% on the net profits he would be rewarded by his efforts. While you of course would save the Company your salary [you] would remain as Chairman of the concern but would not live on the Ranch and by residing in B.C. would be available at any time should Chance wish to consult you . . . .

Cecil had overstepped the mark. In replying in February 1937, Frank abandoned his usual short letter style and took five pages to tell Cecil how wrong he was. Frank described how he had reduced overhead expenses by cutting his own salary, set by his father in 1914 at $5,000. He was paying personally for his car, Nellie Guitteriez's salary, all travelling expenses for ranch and stock association business, and all expenses for entertaining buyers and other businessmen at the ranch. Having never received anything for carrying out the many duties as trustee and executor of his father's will, he had dropped his own director's fee when his codirectors asked for more.

We, Ken and I, have given our best to the family Ranch and you suggest we retire without compensation of any kind.... With only a very small income outside of our Salary, how may we retire?... Did it ever occur to you what Ken has gone without all these years, what every girl has the right to expect, and because she gave Father her word to help me; how few wives have slaved, studied every way of scraping and saving for the Ward family. No woman ever made a greater sacrifice.

He objected to Cecil's remark about the ranch needing a young and energetic man such as Brian Chance for manager. "I'll warrant I have as much energy as Brian." Chance and his wife were receiving $3,000 per annum salary and company travel expenses paid. Frank continued, "Audrey I regret to say has become an invalid. Her trouble is nerves brought about by not being contented with life on the Ranch. Considers we all work too hard, especially Brian.... Compared with our early days, he lives a life of ease."

Douglas Lake's manager of 27 years then justified the presence of each of his "expensive" farm foremen. George Leith at the Home Ranch "will not retire gracefully" and, Frank admitted, he was becoming an expense. Lawrence Graham at Chapperon was receiving "a salary of sixty dollars a month and board. In early and more prosperous days Graham was paid one hundred dollars a month and found." Talbot Bond at Norfolk was receiving the same, and Ward reasoned the case for employing such family men, despite the additional mouths to feed. "You must understand men with families are usually more dependable and make the best foremen. We have experienced other types. They cost us dearly, by stealing stock and sundry things when one's back was turned. To run a place of this size it is absolutely necessary to have dependable foremen in whom you have confidence. When weather conditions are at their worst, they are out and on the job, and not keeping the fire warm, taking it for granted all is well."

Frank's nephew, Curtis Ward, was jointly in charge at Minnie Lake with Jock Hovell at $40 a month and board. "Knowing the difficulty in finding men for these positions... it is well to leave well alone."

Explaining the rest of his staff, Frank wrote:

Storekeepers and bookkeepers, difficult and trying people—our present man as good as any. The store is a big asset. Too cheap a man may be very expensive by helping himself. A good deal of ready money passes through his hands in twelve months. It would be very difficult to detect.

The general ranch hands are paid twenty five dollars a month and board; cooks, $25 to $40. These latter are looking for more pay. As things improve in the big centres they can find work at bigger wages.

On top of wages, there were many other expenses.

For the privilege of grazing over the Crown range, we pay 5¢ per head per month. We lease also from private parties, 53,000 acres at a cost of $2000 a year. Should we lose these private leases, we must reduce our cattle accordingly. To maintain the present herd it is necessary to feed everything through the winter, which means ten thousand tons of hay in any ordinary winter, but one like we are passing through, it requires a great deal more, and though we can ill afford it, we are buying now, to save our cattle.

You have never suffered the agony of mind to find your feed fast disappearing, deep snow, no grass in sight, and the temperature at twenty to thirty degrees below zero; one Station saying they can feed just another week, and so on; scouring the country for feed, driving the stock miles through deep snow; all the time piling up expense and the cattle going downhill.... For six solid weeks the thermometer stood from ten to forty below zero, with a deep snow and now crusted. If you ever saw Russell's picture, The Last of Five Thousand Head, you will know what a nightmare I can have.

To get back to my subject of providing hay, grain and roots, we increased our farming operations, and must go further. To do this we had to create reservoirs by building dams, from these, ditches and flumes all of which must be maintained. At Minnie Lake we have expanded and have a large area under cultivation. At present we [are] purchasing some two thousand tons of hay annually.... This money spent on our property will eventually give us our required supply, and also enhance the value of our lands.

Land Taxes at present $4,000 per annum. Grasshopper tax varies according to the plague of the year, $250 last. Water rates $175. Grazing dues to the Crown $1208. Fence maintenance, trail cutting over these vast areas very costly; in earlier years we did not need or have. Fire fighting, forest fires: in years gone by the Crown fought the fires. Now you must, or lose your grass and fences. The Crown only fights where the timber is of merchantable value. With our woods full of fallen timber killed by the timber beetle, the dead falls are impossible to get through. It then is necessary to cut trail so the cattle may go to where there is feed.

While such expenses had steadily risen, income had fallen off in two areas particularly.

In the year 1918, cattle prices reached the highest price in our history of ranching in Western Canada. The Canadian Government became alarmed at the numbers of females being slaughtered for beef, and so brought about the price spread, from ½¢ per lb live weight from time immemorial, till now the spread is 2¼¢ to 3¢ per lb live weight.... This has meant a continued loss to us of $25,000 per year from our beef sales, which [now] consist of 1000 head of Cows and 1500 steers....

Our horse sales, since motors took the horses' place, fell from $50,000 a year to a few hundred dollars. This sent us out of the horse business commercially....

You write as though speaking for the family, for these are your words, 'We all feel that you have worked so very hard and for such a long time that it is only fair to you and your family that you should take life easier.' These are your thoughts, gracefully put, to make available my salary for the profit and loss account, so you may benefit. Cecil, I am ashamed of you.

Frank tried to explain the economics of ranching to his brother.

Competition does not enter into the cattle business in cost of production in British Columbia. We import our shortage. Price merely governs. It behoves us to produce as cheaply as possible to enable us to make a profit. That is our only competition.

I am trying to point out your folly, attempting to be sentimental in the matter of business, which is selling out a business doomed as civilization over takes it. The longer it continues the less its commercial value....

Look at the cattle industry of the world: England our best market is bonusing her beef producers; Argentine doing the same to her producers so that they may pay the British tariff and get in on an even footing; United States has a tariff of 3¢ and 2¢ against us, we

144

are permitted to sell them 150,000 at two cents a pound duty. We have a surplus of between three hundred thousand and four hundred thousand head to market. Where? Until we find a market that will care for our surplus, how may we prosper?

I am thankful to have in writing from the majority of our shareholders their approval of Mr. Duncan's thesis . . . realizing the serious position of so huge an investment as this.

The English shareholders, inevitably headed by Cecil Ward, replied to Duncan's November 1936 dispatch in a joint document, pleading to raise the sale price to a bedrock figure of $750,000. In 1937 that kind of money was not available, but as the worst of the '30s passed and economic conditions improved, the asking price rose once more.

But in one way, Frank Ward did heed his brother's words. In August 1938, after yet another bad winter which was so perverse that even the horses had to be fed hay, he formally agreed to retire from the management of Douglas Lake Cattle Company on 30 April 1940 and to let Brian Chance take over. Ward would then be two months over 65. In fact, Chance took over a great deal of the active management of the ranch from the day Ward signed the document.

Though Frank was around the ranch far less from the late '30s until his official retirement date, he still kept a tight control on the reins by writing frequent notes. He continually instilled the doctrine "Take care of the pennies and the pounds will take care of themselves" into his understudy, as is evident from these messages sent to Brian in the late '30s.

"Now the cowboys are away, it surely won't be necessary to have a dishwasher . . . . Is there any need for Joe & Tommy to *both* have crews? A dollar saved is a dollar made."

"Please look carefully over the time book & cut down as much as you possibly can on our help. . . . In other words, get rid of everyone we can manage without—our overhead will eat up all our benefits from the improved prices."

"I cannot understand why? Four Chinamen in the Ranch house—not one of them told to keep the Ranch house in order—sitting-room, upstairs. This must be cared for. There are berries, currants to pick. Lim Louie & the boy should [see] to these things till they are wanted elsewhere. It's a mistake to take a line of least resistance."

Experience had made Ward wary of the Chinese on the ranch. One Home Ranch cook who had stayed so long he was almost indispensable was fired for making prune whiskey and bootlegging it to the farm crew. Another Chinese cook went after Betty Ward with a knife. Yet another peddled opium until Ward caught him.

In 1936 Ward had advised, "Keep in mind that we mustn't spend a dollar, which means we must *not* have anyone we can manage without on any of the 'Ranches' or 'Stations.' It was alright to allow old Joe to have 2 bags of spuds, but don't forget any we can spare are worth money & will sell for a good price."

And a year later: "Re—Yard fence. At present this I considered an unnecessary expense. I should have laid off those two men & so saved a few dollars we didn't need to spend. Always keep in mind to live within our income or you'll see the company go broke in a short time."

Frank had learned good management practices through trial and error and passed some of his theories on to Chance. "You'll find if you don't guard the spring grass very carefully, you'll regret it in days to come. Every horse counts. They are the devil, as I found to my sorrow, doing but as you [are doing], till at last I shot [sent] them in the timber to live or die in desperation, believing it better to lose the horses rather than the grass."

Again on spring grass: "I hope you are having a good winter and that you will not graze too long. Remember the grass is worth double in the Spring, when the cattle refuse to eat the hay and want to wander." And in the middle of one mild January, "I hope you are feeding most everything. It's poor policy to hold off too long. Your cattle will suffer in April & May. You save hay at the expense of your cattle & grass."

On his wintering policies: "If the snow goes & the weather opens up, you will be safe. If not you must immediately start figuring to make your feed hold out till the end of March. This may mean cutting your steers' ration in half. Better to let them go than lose any of the other cattle. The steers will live but will of course be late beef, that can't be helped. You may have to get hay in, by rail from Chilliwack or Fraser Valley.

Should immediately be getting quotations F.O.B. shipping point."

But when the grass was there, Frank wanted it used. "Coming through Aspen Grove today, I notice *Guichon* has a *bunch* of *cows* down the one mile, in *grass* up to *their knees*—opposite the Skookum Illahee. *I didn't see any of ours.* The grass is splendid all the way."

Apart from leaving the ranch to attend the Kamloops, the Calgary and sometimes the Lacombe Bull Sales, Frank still managed to get away for polo. Since his early days in Alberta, he had never tired of the sport and had gained a handicap of two, making him a most useful addition to any team. For a time in the '20s, he raised polo ponies at Douglas Lake, and Ralph Chetwynd and Tommy Wilmot, a four-handicap player, trained these ponies at the Brush Camp. From the '30s on Frank had personally paid for transporting the 12 polo ponies needed for a match, regardless of which team he was on—Douglas Lake, Kamloops, or Westwold—or of the distance to the game.

He played his last game aged 65 on a 35-year-old horse, an event the newspapers noted. That last game finished polo in Kamloops, for without Ward's leadership the club disbanded. His polo ponies ended their days in lavish comfort at Douglas Lake.

Then as president of the B.C. Beef Cattle Growers' Association Frank attended association meetings, lobbied government ministers and sat on boards such as the B.C. Beef Marketing Board, although he opposed control of beef marketing by the board, believing that it would take away the freedom and rights of the ranchers. He sent letters to many newspaper editors and began a huge newspaper debate over its good and evil. The beef marketing board eventually held control in the '30s, but not thereafter.

Later in the '30s, the grading of beef received a great deal of attention. Frank Ward was in favour of improving the old act, arguing that if the government could grade properly, the producer selling good stock would receive more while the producers of inferior animals would receive less.

At first the packers and then the housewives were against

the scheme. With Canada producing surplus beef, the Americans had agreed to import a certain number of cattle annually, starting in January. Until this quota was filled, sometime in the summer, Canadian prices were higher by as much as a dollar a hundredweight. The Grading Act came in at the end of 1938, just when prices rose, and though the price increase happened every new year, housewives blamed the high prices on the Grading Act.

A beef advertising campaign of the cattlemen's association neatly coincided with the Housewives League's boycott of beef. Beef won. Meanwhile, Frank Ward was lobbying the minister of agriculture to put more teeth into the act and provide tighter controls. The new grading policy became accepted—benefitting at first consumers and later producers. This was a suitable feather in Frank's cap for his last year of presidential office.

Frank's days at Douglas Lake were slipping by, and as the time came for moving from his home he might have been expected to feel closer to the woman with whom he had shared 30 years of trials and tribulations. But now, her spirit broken, Kenny no longer wished to live out her days with the person who had directly and indirectly caused her such a hard life. Perhaps she believed that during all these years, Frank's first love had been for Douglas Lake Cattle Company and not for her. Kenny moved to the Okanagan and later to Ontario, where she outlived her husband.

The man who considered that his greatest accomplishment in running Douglas Lake was to have kept it solvent retired to Victoria. He wrote Chance, "Good luck, Brian, and all success to your management. Watch your *loans* & *if possible* never borrow more than 50% of your borrowing power except in an emergency like a hard winter. In keeping to that principal [sic] you'll make a success & dividends to please all." And later: "Don't give a damn for any criticism, also think '*big*.'"

CHAPTER THIRTEEN

The Company ... farms some 4000 acres under irrigation, puts up
6000 tons of hay, threshes 200 tons of grain ... employs 60 men
the year round and increases to 180 men for haying.

Francis B. Ward, "Douglas Lake,"
undated

When Frank Ward became manager, Douglas Lake was
growing roughly 2,000 tons of hay annually from 1,000 irrigable acres at Big Meadow, Brushy Field, and a number of
natural hay meadows at the Home Ranch, Chapperon and
Minnie Lake. Three times that figure were needed each
winter, however, and Ward decided to step up the farming
operation. He gradually gathered good farming men around
him to take charge at each hay ranch.

Of the farm foremen that Ward inherited from Greaves's
day, only Jock Hovell stayed on; he became one of the earliest
foremen at Minnie Lake when that hay ranch 16 miles southwest of the Home Ranch was putting up 250 tons annually.
He was a good farmer of the old style and handled his large
crew capably, although on more than one occasion he en-
dangered his position by the troublesome women he brought
to Douglas Lake.

In 1911, Will Oliver approached Frank Ward to find a job
for George Hector Leith, a Scot who, after helping to repair
the declining fortunes of his titled family with Yukon River
gold, had entered a partnership with Will in a salmon cannery
that had since gone broke. Ward gladly added Leith to the
hay crew. A strong-willed man, Leith soon ran afoul of the
studman, Tom Stewart. A fight seemed the only way to solve
their argument, and Leith came out the victor. Stewart left
the ranch shortly after.

Frank Ward then put Leith in charge of the Home Ranch

farming operation, handling the biggest haying crews on the ranch: up to 70 men and women, half of them living in tents scattered over the Home Ranch yard. Leith had much to contend with—including Ward's habit of firing any slackers in sight, without first consulting his foreman.

Chinese irrigators had, under Greaves's direction, built ditches from the Nicola River to water some of the land above English Bridge, three miles from the Home Ranch. Now, Leith had the Home Ranch crew tackle 200 adjacent acres which, when cleared, levelled, ditched and seeded, became the productive New Ground.

In 1920, Lawrence Graham, a 32-year-old Lancashire emigrant, became foreman at Chapperon. As well as having worked for Leith, he had farmed his father's homestead in Alberta, driven transfer and dray, and worked as a carpenter on the building of Victoria's Empress Hotel. Graham's first job at Chapperon was to make the old bunkhouse a home for his family, and he relocated it by pulling it with teams over log rollers to stand amidst tall trees not far from the north end of Chapperon Lake.

Chapperon hay ranch was still largely willow brush, which Graham and his crew began clearing. Lawrence's wife, Lillie, took on the job of cooking for her husband's crew: peeling root vegetables, churning butter, butchering quarters of meat, and baking bread and pies seven days a week, rising early and retiring after putting away the supper dishes. In the height of summer when more than 30 temporary workers expanded Lawrence's hungry crew, Lillie would find herself a helper; one such was Whispering Julia, an Indian woman whose larynx had been permanently injured by a kick from a horse.

Lillie Graham did not meet Mrs. Frank Ward, the only other white woman on the ranch, until she had lived at Chapperon over eight months, for an 11-mile wagon trip was not easily fitted into her daily chores. They eventually met when Lillie's leg was bitten by an insect and swelled so alarmingly that Mrs. Ward rode over to apply hot packs to bring the swelling down.

A third white woman came to live at Douglas Lake a month

or so later, in December 1921. Earlier that year, Talbot Bond and Robert Moir had sold their Norfolk Ranch to Douglas Lake Cattle Company. Bond, a tall, dark-haired Englishman of the most conservative habits—except for his chain smoking—was asked by Frank Ward to act as interim farm foreman. Bond worried that life would be lonely without his partner, Moir, but agreed to stay on. Soon after, he stopped in at the Ranch House, and Leith asked, "Why the hell didn't you marry Constance Johnson?," referring to a lady who had journeyed out from England as a companion to Mrs. Moir but had left for Australia some time later.

"I did," replied Bond. "She's out in the car."

His wedding during a short trip to Victoria was the most impulsive move he ever made. As Bond's interim position as farm foreman of Norfolk hay ranch stretched from months into years, Ward came to value Bond's steadiness, which went hand in glove with his resistance to change.

The melting of the snows, the thawing of the lakes, the return of the wild ducks and the first new shoots of bunchgrass heralded the time when the herds vacated the winter feedgrounds, and Jock Hovell, George Hector Leith, Lawrence Graham and Talbot Bond could begin their spring work on unproductive hayfields: plowing up, disking and harrowing, and reseeding to tame grasses and legumes those fields that had come to the end of their ten-year productive cycle. Towards the end of March, up to 20 Chinese irrigators arrived to burn and clear the old grass blocking the irrigation ditches. Then they would flood-irrigate the haylands, and in a few weeks seasonal haying crews would flock to Douglas Lake, swelling the total payroll to 180 men and woman.

Through the hot, mosquito-ridden summer weeks the mowing, raking, cocking and stacking crews busily harvested. The Clydesdale teams were now gentler than when they had first come off the winter range, and with them a good skinner (teamster) each day left swaths of forage in many huge concentric circles and ovals to be swept together by the dump rakes. For years, pitchforks were used to toss the loose hay onto wagons which were then pulled to the haystack by horses. Later, the bull rakes, or sweeps, partially took over this job. A two-horse team pushed these three-wheeled rakes forward until the rake teeth, which protruded seven feet in front, slid under the sun-dried hay to gather a load of just under a ton. The operator pulled a lever to raise the teeth with their load of hay, and directed the horses into the stack. Here the hay was lowered onto slings that were then lifted with a pulley arrangement, the sling ropes being pulled by other horses until the hay could be dropped into position on the top of the stack.

Various types of stackers were used: the A-frame stacker, which Talbot Bond liked best; the jammer or swinging boom stacker, which Lawrence Graham favoured; the gin pole stacker, which Leith preferred to use, and others besides. In swampy areas, stoneboats—known locally as sloops—functioned best. A set of slings spread out on the flat platform of the sloop soon disappeared under a mound of pitched hay. At the stack, the slings hooked onto cables for stacking.

The Home Ranch has the lowest elevation of the four haying areas of Douglas Lake, but even there at 2,600 feet, the growing season for two hay crops is short. Higher up at Chapperon and Minnie Lake, only a very light second crop of alfalfa can grow on the sidehills; at Norfolk, only one crop can reach maturity. If the hay were all cut and stacked by the end of September, haying had gone well.

Fall was the time to harvest the grain. In 1910, Greaves had bought an old steam engine which powered a threshing machine drawn by a four-horse team. Before frost hardened the ground, plowing for the next year's seeding could begin. Whitewashing yard buildings and cutting firewood for winter were other fall chores.

Regardless of the weather, 15 December was Frank Ward's day to bring in the weaned calves from the range. The rest of the herd stayed out to "rustle" the bunchgrass as late as possible. Any time between Christmas and late January, depending on the snow, riders divided the cattle between the different hay ranches for feeding with team and wagon or team and sleigh. A half ton to a ton of hay per head brought the herd through the three winter months; the dry cows and older steers weighing around 1,000 pounds apiece consumed the smallest hay rations and would look like greyhounds in the spring.

The monotony of doing nothing but feeding cattle twice a day for three solid months, often in below-zero Fahrenheit temperatures, was the lot of Douglas Lake's smaller winter crews. Winter was when the farmer, cowboy or cook found out if he were susceptible to becoming "bushed"—deranged by loneliness. The only variety to the winter was provided by the heavy work of cutting and hauling hundreds of 200-pound ice blocks to the icehouses that adjoined each hay ranch's meat house. Four feet of sawdust on all sides effectively insulated the ice for months. In summer, several blocks were skidded as needed onto a strong, ventilated shelf above the meat and other victuals that needed refrigeration.

At Chapperon, Lawrence Graham continually managed to improve the acreage under crop and to enlarge the irrigable acreage. Unlike the other three foremen, all much older than he and more set in their ways, Graham was always ready to try different methods. He handled his men exceptionally well, getting the best out of everyone in his crew.

As at the Home Ranch, Chapperon had its New Ground, 250 acres of sidehill that Graham's crew first turned over, using gang plows, in 1929. More acreage expanded the haylands until eventually Lawrence built up Chapperon's hay crop from 400 tons to over 2,000 annually.

When Frank Ward was making savings wherever possible in the '30s, he despaired of some of George Leith's farming methods. "Why? oh Why? are you running the tractor without the harrows behind the discs? You will now have to go over all that ground again costing God knows what in gas & time. . . . What is the use of a tractor if not to save time and money[?]" Having Ward's eyes on him all the time was a disadvantage of being the Home Ranch farm foreman.

Up at Minnie Lake in the early '30s, Archie Davis, an Aspen Grove stockman, temporarily replaced Jock Hovell. Davis's wife could handle the cooking job during most of the year, but—as Lillie Graham had found earlier—she needed another pair of hands in the kitchen during the eight to ten weeks of harvesting. A young Indian girl took this job, but she left quickly, and so did her successors. The word spread as fast as a grass fire that the dishwashing job at Minnie Lake was a perilous one, for Archie Davis occasionally got drunk and showed his haying crew how strong he was by catching his wife's little dishwashers and holding them upside down by their ankles.

Up at Chapperon, Lawrence Graham worked his crew efficiently, around the clock and around the calendar, so that the occasions were few when Ward had to travel there in the ranch Model A Ford pickup; his rare presence always frightened the Graham children into hiding. In winter, Graham organized his men so that a couple of teams finished feeding the cattle early enough to go up nearby Cayuse Mountain and haul back a load of logs before sunset. By the end of winter the pile of logs was high, and when spring work was over there was just enough time before haying started to process the logs into lumber.

The old steam engine, when not threshing grain, powered the sawmill at Chapperon. The authorities required that only a man with third class engineering papers might operate it, and Leith alone possessed such papers; despite this, Graham occasionally ran it. Many years they cut and used 250,000 board feet. Most of the lumber went for building irrigation flumes and the rest for houses, sheds, barns, yard fences and bridges. Ranch lumber from the Chapperon sawmill had built most of the buildings on Douglas Lake—and that was no small number.

Not long before Lawrence came to Chapperon, Kenny Ward had tried to cure bacon. She bought pickling spices and had a log as big as an Indian canoe hollowed out to house the bacons while they cured, but her entire batch went bad. Sometime in 1923 the spices, hollowed log and half a dozen hogs arrived at Lawrence's door. The hogs fattened well on the cookhouse slops, grain and whatever they could root around the yard. By fall they were ready for slaughter.

Live pig to cured bacon is a long process, one that Lawrence had learned during his time at the Guichon Ranch. First he killed, bled and scalded the pigs, then he halved them and hung them to cool. After cutting them into bacons, hams,

Boston butts or picnic shoulders, he placed the portions in a strong mixture of saltpetre, sugar, water and spices. Lawrence then injected the brine into the centre of the meat with a big syringe. Once the bacons and hams had cured, he smoked them in the smokehouse over an alder fire, then wrapped them in brown paper and put them in cheesecloth bags that Lillie had sewn. Finally he dipped them into a barrel of a lime and water solution which made an airtight covering. He was so successful at curing that Frank Ward set up a proper piggery, with sows and a boar, at Chapperon. Lawrence designated one of the non-haying months for pig killing, and it became a big affair requiring many men.

In winter, he cut back the boar's tusks. A male pig, which can grow to a size of 700 pounds, is a powerful animal with strong jaws, lethal tusks and often a mean disposition. Lawrence devised a boar-wrestling technique: he took the animal to a slick area of the frozen lake, where its hooves would make it lose its footing. Once down, the boar was not nearly so dangerous and Lawrence could move in and cut back the tusks.

The economics of raising hogs soon became clear: each sow must wean five to cover overhead; three more in a litter would turn a profit. Not everyone could make money on pigs, but Lawrence did. Chapperon supplied the entire ranch with bacons and hams; any extra found buyers in town. None of the pork was wasted, and as well as pork sausages, Lawrence made a superb liver sausage. His head cheese and pressed tongue were delicacies, and his wife's pickled pigs' feet were better than any sold in a store.

Lawrence Graham later had to turn the bacon shop into a schoolroom where his four children could take lessons from a tutor hired by Frank Ward; a tutor was also sent to Norfolk for Talbot Bond's daughter, Rosalind. When the three Graham boys reached their early teens they signed on with their father's crew.

Douglas Lake started showing cattle in the mid-'30s, and Graham's youngest boy, Raymond, helped by feeding some of these show steers after school. The 70 head of 1,100-pound two-year-olds picked out of the commercial herd in October were penned at Chapperon. Their daily ration was 15 pounds of grain so that when they were broken to halter for showing at the Kamloops Bull Sale and Fat Stock Show in March, they weighed around 1,500 pounds.

Getting the prime cattle to the show was a major problem. Their superb condition would suffer if the cowboys drove them in, so two days were devoted to transporting them by truck over the rough 70-mile route. Two four-horse teams preceded the trucks to the first bad mud hole, and if the cattle trucks got stuck, the horses would pull them out. Then the trucks would drive on until they got stuck again and had to wait until the teams arrived to pull them out once more.

Douglas Lake was up against fierce competition from the Guichon Ranch, the Gang Ranch and the Eldorado Ranch to start with, but as time went on its animals began to win prizes for the Grand Champion Steer, the Reserve Champion Steer, the pens of 5 and the carload lots of 15.

Raymond Graham started working full time in 1939, the year that new bunchgrass sidehill land at Harry's Crossing, east of Chapperon, first became ditched for irrigation. Early in the spring, his brother Ralph had moved up there with two other farmhands, a cook, and a team and wagon loaded with tents and supplies. Raymond joined them later that spring. Together the men used team and slip scraper (an earth mover) to finish the 100-acre-foot dam, and then plowed up and seeded 100 acres for that first year's hay crop.

Chapperon men cleared willow brush, picked rocks, built a stackyard, erected permanent buildings and plowed up more land over the years until 550 acres were under cultivation. Run as an offshoot of Chapperon and overseen by Lawrence Graham, Harry's Crossing housed a separate crew that stayed all summer, until they had stacked the hay. In winter the cowboys drove a separate herd there, feeders and one cowboy staying all winter to feed and care for the cattle.

George Leith, the foreman who had put up the Home Ranch hay for almost 30 years, retired in May 1939. Retired, Leith was as loyal as ever to Douglas Lake, returning in 1942, when World War II made labour short, to help Jock Hovell put up the Minnie Lake hay. Before his death in the late '40s,

Leith often accompanied Frank Ward back to Douglas Lake
for the pheasant- and duck-shooting seasons.

During Ward's management, George Hector Leith was
foreman at the Home Ranch for almost 30 years; Jock Hovell
at Minnie Lake for over 20 years; Lawrence Graham at
Chapperon for 20 years; Talbot Bond at Norfolk for 19 years.
And Hovell, Graham and Bond—the unsung heroes of the
ranching world—had more years of farming at Douglas Lake
ahead.

By 1940, when Frank Ward retired, his program of
increasing hay production had yielded tremendous results. In
1910 the hay crop had weighed 2,000 tons. In 1932 it reached
6,000 tons; as well, 16,000 bushels of grain were being grown
to finish cattle and to feed horses and pigs. Hay production
continued to rise until 1938, when Douglas Lake farmers cut
and stacked close to 8,000 tons from the 4,000 acres of hayfields. In a winter without prolonged snow or exceptionally
severe cold, 8,000 tons of hay could bring the 10,000-head
herd through until spring. By once producing such a crop,
these farmers had proved that it was possible for Douglas
Lake to become self-sufficient in its production of hay.

CHAPTER FOURTEEN

With all his failings he was a tower of strength to me especially in my early days, and I thank God he lived through my reign at Douglas Lake for he made it that much more simple... enabled me to get away from time to time and know that the cattle at any rate would be properly looked after. I never went off for a long voyage without having old Joe promise me he would not get off on a drunk during my absence, and to my knowledge, he never did.

Frank B. Ward to Brian K. de P. Chance,
7 June 1945

"Light the branding fire!" boomed Joe Coutlee, true to his nickname of Roaring Bill. He was a muscular man of over 200 pounds, and although he was not quite 6 feet tall, his demeanour led his crew to believe that he was much bigger. It was the end of another hot June day, not the usual time to start a branding fire, but soon the coals were glowing red. "Put an iron in the fire!" cried Joe. A cowboy grabbed one of the irons used to brand the Douglas Lake stock with | | | on the right hip, and pushed it into the hot coals.

"Bring my wife here and tie her up, boys. I'm going to brand her so everyone will know whose she is." A cowboy threw a rope around the terrified woman and dragged her to the fire. Joe lifted her skirts and, ignoring her screams and the muttering of the incredulous men, pressed the red-hot iron to her bared hip three times.

When Frank Ward had first taken over the management of the ranch he told Joe Coutlee that he could not have more than one woman living with him in the cow camp. So Joe chose a very good-looking young Indian woman whom everyone called Muggins, though she insisted that her name was Mary Ann Horne Jackson Coutlee. She was the unfortunate woman he branded, and despite this and other cruelties, Muggins stayed with Joe for the rest of her days, caring for their six children. Most of the time, Joe and Muggins resembled any other happily married couple, but when they got on a drunk, anything could happen. All in all they were a well-matched pair.

Joe's French Canadian father, Alexander Coutlie, had left his home in Trois Rivières, Quebec, for the California gold rush. Later he joined the Fraser River rush, but instead of mining again, he went into the hotel business at Yale and Boston Bar, where his menu was said to include cat meat stew. In 1873 Alexander and his Thompson Indian wife preempted land on the Lower Nicola River, and operated a flourishing hotel and a store there, as well as a small cattle ranch.

Young Joe's practical education included working in his father's saloon and helping to capture the McLean Gang. He received proper schooling at Columbia College, New Westminster.

Once, as Joe's school holidays were coming to an end, he and his father drove some cattle to Yale together, arriving a few days ahead of the sternwheeler. There they were joined by J.B. Greaves with more cattle, and Benjamin Van Volkenburgh with some sheep. Their journey downstream to New Westminster went well until the boat grounded on a craggy rock, tilting badly. The captain ordered Van Volkenburgh to put all his sheep overboard to lighten the load, intending to pick them up once he floated the boat off the rock. Unfortunately, the sheep did not swim to shore but swam in a circle until they were exhausted; all drowned.

With the sheep gone, the captain was able to right the boat and the journey continued uneventfully to New Westminster—but here was more trouble. The buyers of the area would not give Coutlie and Greaves the price they were asking for their cattle. They could not afford to take the stock home, so Greaves obtained some pasture and the two killed and dressed the animals themselves; they got their price by selling the meat in carcass form direct to the butchers.

By the age of 23, Joe Coutlee had ridden with some of British Columbia's best cowmen, when he signed on with J.B. Greaves and his top cattleman, Joe Payne. In those days of the West, a cowboy was an all-round man of the range. He could ride any horse, mean or gentle; break, shoe, pack and care for a horse. He was proud of his rigging and knew how to repair it. He could handle a rope and knew all about cattle. He rode strange country, in hills or in timber, without getting lost. He could build or fix fence, and handle a gun. He cooked for himself and slept under the stars. By the time Coutlee became Douglas Lake's cowboss in 1896, he was all of this, and he expected his 20 riders to measure up, too. Those who did not rolled up their blankets and found other work.

As head cattleman for Douglas Lake Cattle Company for the next 49 years, Joe Coutlee had a down-to-earth approach and latterly a quiet dignity that endeared him to whites and Indians alike—the Indians holding him in near reverence. It was a proud man who could say, "I rode for Coutlee."

When Francis B. Ward became manager, Joe already had 19 years in the saddle at Douglas Lake, and Ward so respected his cowboss's knowledge of cattle that he resigned himself to Coutlee's habit of going on a drunk once the work was over.

In Greaves's day, anyone using his own horses at Douglas Lake had been paid extra. Frank Ward observed that privately owned horses often came into camp sore-footed and in poor shape and left a few weeks later shod all round and in sleek condition, having spent little time under the saddle yet having consumed the company's grass. Ward ended this practice. It was in such a way, however, that Coutlee had been able to develop his own remuda on the ranch. He improved the breeding of the original mares obtained from the Indians by using his own stallions. Soon as many as 40 horses wearing Coutlee's JK brand ran on the Crown range around Pothole. Over the years, many of these JKs rode in the cow camp and worked hard, and for one or two seasons a JK stallion bred the company mares because he was the best available. Not until after Coutlee's death many years later was the string sold, several of which, such as Reckless George, Fadeaway and Devil's Dream, became well known in rodeos across western Canada.

Each cowboy rode a personal string of six or seven ranch horses—five fairly gentle and one or two that were being broken. He could use a different horse each day so that none would suffer saddle sores or become sore-footed. Sometimes, though, Coutlee would advise a cowboy to "ride that horse four or five days straight," for a fair number of the remuda remained wild—even treacherous—to the end of their days, and the only answer for them was many miles of hard riding.

There were a number of reasons for the wild strain. J.B. Greaves had never brought in a good stallion to breed the saddle horse brood mares. Frank Ward thought any strong, sound, ranch-raised horse good enough to breed the colts which cowboys would break and spoil. Coutlee was interested more in performance than in appearance or temperament. Thus the Douglas Lake cow camp horses were crossbreeds of Clydesdales, thoroughbreds and cayuses. Because of the Clydesdale influence, some were not clean-legged but feathered, some were soft-boned in the legs, perhaps from a mineral deficiency, and others were pudding-footed, taking a size four shoe.

As well, Coutlee never allowed his cowboys to break a horse under the age of four; in his opinion a younger horse was not strong enough to work. Some reached five or six before they were first handled by the crew. Occasionally the crew left alone a horse which bucked badly when first ridden, and after a year the animal would be even more set in its ways. Some cowboys considered that the best way to break a "wild" was to corral it, rope the front feet, get the horse down, cinch up the rigging, mount and stay on once the animal got up, and ride. Others believed that this spoiled a horse forever.

So when stampedes became popular with the public, and rodeo hands came out to the ranch for a little practice before the circuit started, there was never any problem in finding them a rough string of mean, spoiled, hard-mouthed horses that no one else was anxious to ride. These bad horses would "shake hands with their rider" (kick with a front hoof) as the halter went on, or swing away as he climbed up, or put their heads down while galloping towards a low branch, or cow-kick as he climbed down. They were always ready to catch their rider off guard.

At the end of each day, a cowboy would run his wilds into a corral so he could halter-break the new ones in his string.

Each was taken through one new exercise, on the average, every time. Holding the end of a rope tied with a berry hitch around his horse's neck and head, which pressed on nerves to make it pay attention, the cowboy soothed his mount with soft chatter. Tying the rope to the snubbing post in the centre of the breaking corral, the cowboy "sacked out" his horse, flapping a piece of cloth around the nervous animal's legs, shoulders, neck and head to get it used to the unexpected, and then picked up the horse's feet to prepare it for shoeing. Next came the saddle alone and then the rider, and the new horse was ready for the work that would wear more rough edges from its disposition. Coutlee demanded that all wilds be taught to stand when caught and to be led out of the corral or yard when haltered up. The rider undertook the rest of the breaking in his own time.

By the beginning of January each year, Douglas Lake's herds would be in their wintering positions—the weaned calves on feed and the older cattle, if not already on feed because of snow covering the bunchgrass, staying on the winter ranges close to the ranches where feed was available when needed. During the winter months, Coutlee's crew would complete the weaning. They had taken the biggest calves from their mothers in November, but as a cow/calf pair consumes less hay before weaning than afterwards, they left many as late as possible. Then they cut out any slickears, those calves born too late for summer branding, feeding them together and branding them during any warm winter spell or early in spring. At least once every winter month, Coutlee and his crew checked each herd, whether it was grazing out or eating hay, and cut out the tailenders, putting the thin and sickly into hospital pens for closer attention and a larger daily hay ration. There might be the odd drive of last year's fat cattle taken from the Morton to the CPR station at Nicola.

The cattle that were to winter feed at Westwold, if they had not gone already, went down to be overseen by Tottie Clemitson or, at times during the '20s and '30s, by Frank Gordon. He was a good friend of Frank Ward having played polo with him on many occasions; a small man with a fierce handlebar moustache who could have posed as a Yeats Brown Bengal

Lancer. Each year when the Douglas Lake cowboss arrived
with the drive at Westwold, Gordon's greeting was the same:
"Hello Coutlee, my dear chap. Glad you've arrived safely.
Hope things at Douglas Lake are going strong. How's my pal
Ward?" And that became the crew's nickname for Gordon:
"my pal Ward."

On these seasonal trips to Westwold the cowboys revelled
on liquor bought either from the bootlegger living on the Salmon River canyon road or from the bars in the Westwold or
Falkland Hotels. The second day of those drives was often
chaotic as a result. Once a Spahomin Indian, Joseph Frank,
who had drunk quite a bit, fell off his horse. Deafie, as Frank
was called because of his hearing problem, had been riding in
the middle of the 600-head herd and lay in a stupor on the
winding canyon trail. The cattle immediately behind stopped
to stare at the strange spectacle of a man lying on the ground;
the rest of the herd kept coming. The uphill side of the bank
was too steep for the crowding cattle to climb, so they were
pushed down the perpendicular sidehill to the Salmon River
below.

The first cowboy to round the corner, Seymour Williams,
quickly hitched a rope around Deafie, tied the other end to a
tree, and lowered him down the slope where he was out of
sight and could no longer upset the cattle. But 92 head had
fallen from the trail into the canyon below, more than 30 of
them losing their lives. The men could not even salvage the
hides, for they had no time to skin the dead cattle until the
next day on their return trip and by then the carcasses had
frozen solid.

The years before the CNR came through, when land was
unfenced, were the mud hole years for Tottie Clemitson. He
would ride his best pulling horse, Toby, from mud hole to
mud hole all along the ten miles between Westwold and Falkland, where the Douglas Lake cattle were feeding: Tottie and
Toby set a record one day by pulling 27 head out of potential
graves.

At times, to relieve winter's monotony at Douglas Lake,
Coutlee and his crew would seek diversion. Once, after the
whiskey had been flowing for two days at the Morton,

Coutlee climbed aboard the chuck wagon that Whistling Tex Turner was driving to the Home Ranch. Turner's patter and his display of whip-cracking irritated Coutlee, who decided that he would show this Texan how a Canadian skinned a team. Coutlee stood up, threw down the lines and began to whip the horses into a dead run. The inevitable happened: the horses galloped wildly into a gulch, killing one and smashing wagon and contents. One of Coutlee's ears had to be sewn back on. Turner left the next day; if this was how they operated in Canada, it was no place for him.

The cowboy crew dehorned in March, both at Chapperon and at the Home Ranch, where the yearling steers and heifers were wintering. Dehorning was the messiest and most disliked job of the year. It was done mainly because the government charged ranchers a dollar per head for every animal shipped with horns, on account of the damage done to live meat in transit. For many years, the only tool that dehorned efficiently was the Keystone Dehorner—a pair of long-bladed shears. The aim of this brutal instrument was to cut the horn off a quarter inch below the hair so that it could not regrow. The stump bled profusely, and a variety of liquids and powders that speeded blood clotting came into use. If the animal in the chute moved at the wrong time, so that a little more than just horn came away, or if the blood coagulant was ineffective for any reason, the yearling might bleed to death.

Turnout to spring range often coincided with the arrival of the first calves, and by April, when the hills were starting to green up, calves frolicked everywhere. During this peak growing season of the delicate bunchgrass, Coutlee was constantly moving the herd to higher elevations and the less susceptible timber grasses, a job made harder by the cattle's preference for the feed they were leaving over the feed they were going to.

During April 1930, soon after the cowboys had started moving the cattle to spring range, Coutlee noticed that at least 100 yearling steers in a field of 900 showed signs of paralysis. The next day, Eric Hearle, the entomologist at the Livestock Insect Laboratory in Kamloops, diagnosed at the site that the cattle were hosts to a plague of adult blood-feeding Rocky

Mountain wood ticks. He estimated that some animals harboured as many as 150 ticks on the backs of their heads and between their shoulders. Following the entomologist's directions, Coutlee's men began scraping the ticks off the cattle with currycombs or their bare hands. They applied a mixture of raw linseed oil and pine tar to kill the more stubborn ticks. Some of the cattle jumped to their feet immediately, others rose after some time had lapsed, but 50 of the young steers died from pulmonary paralysis. Coutlee's men had to be wary, for humans are as susceptible to these ticks as are cattle or sheep. This was the first major outbreak of tick paralysis in cattle within British Columbia, the province shown to have the highest incidence of such paralysis.

From then on, during the tick months of April and May, Ward and Coutlee either kept the ranch stock away from the tick-infested areas of the ranch—open, rocky, southern exposure bunchgrass range such as the Howse Field and the Gilmore—or daily checked the cattle in such fields, watching for the least sign of paralysis and praying for the first hot day that would chase the ticks away for another year.

By May, the spring ranges were reaching their climax growth—time for the cowboys to split the herd and drive the cattle on to summer ranges. Tommy O'Rourke and his crew would drive 4,000 head of yearling steers, yearling heifers, and two-year-old heifers with bulls towards the north end of the ranch, grazing them from Salmon River west to Pennask Lake. The 6,000 wet (milking) stock with their calves and the year's beef stock went south with Coutlee.

Late May and early June was branding time at Douglas Lake's Hamilton Corrals, which lay 14 miles due southeast of the Home Ranch on the northern boundary of the Hamilton Commonage. In many ways this was the big event of the year. For two weeks all the ranches that shared the Hamilton Commonage—Douglas Lake Cattle Company, Guichon Ranch, Lauder Ranch, Abbott Ranch and Sellers Ranch—worked together. Douglas Lake's only chuck wagon bulged with pots, pans, dishes, cutlery, food, tents, bedrolls, branding irons, vaccines, fencing wire, horseshoes and more. Another box wagon carried the overflow. Each ranch's wagons, horse strings and crews set up camp at the Hamilton Corrals, and dry timber—mostly cottonwood—was piled for the branding fires.

Coutlee, it seemed, never slept while camped at the Hamilton Corrals. In the dead of night, he would wake up refreshed and roar, "Wranglers," whereupon the wranglers or jinglers for the day would rouse themselves, mount the wrangle horses and ride out to drive in and corral the rest of the horses. Breakfast was next: beefsteak, bacon, beans, bread and butter and black coffee, served from the tailgate of the chuck wagon and eaten under the canvas fly cookhouse tents by the light of coal oil lanterns and candles. Each man would then identify his horse out of the 150 head in the four corrals, catch it, saddle it and mount, all in the light of the new dawn.

Four thousand head of cows and calves belonging to the different ranches were scattered over the 16,000 commonage acres by June. Coutlee divided the Hamilton into 10 to 12 wedge-shaped pieces and sent riders out to a new piece each day to bring in the 400 or so head found there. The riders would then match mothers and calves and separate them by their brands into the different herds. By the time an early lunch was over, the 30-foot-long branding fire was red-hot and the various branding irons were heating in the coals. The cowboys branded the calves from each ranch as a group, ignoring the cows bawling from the other side of the fence. By brute force, two cowboys wrestled a calf to the dirt and held it down while another branded it and a fourth cowboy notched its ear, castrated it if a bull calf, and inoculated it against blackleg, a bacterial disease. Several teams worked at once, scraping their irons clean of hair and reheating them between calves. When all the calves from one ranch had been branded and neither Joe Coutlee nor Johnny Guichon could find any still slick, Coutlee gave the word to let them out to rejoin their mothers and to start the next bunch.

Once all the men had finished branding, around mid-afternoon, they drove the cows and calves back into the pie-shaped piece of commonage where they had been that morning. If Billy Lauder or his son Joe were there, they kept their branded cattle back, and when a big enough group had formed, they took them to summer range north of Douglas Lake's Home Ranch. Sellers, and Abbott's foreman, November Gottfriedson, did the same with their respective herds.

By the end of the '30s, grasshoppers had eaten off the Hamilton Commonage and the cattle had overgrazed it so badly that it was fast becoming a dust bowl. Douglas Lake suggested dividing it up. Jointly with Lawrence Guichon, Brian Chance wrote a report, and the B.C. Department of Lands allowed the permittees to subdivide the old commonage into three areas, one for Douglas Lake, one for Guichon Ranch, and one for Lauder and Abbott to share. Sellers had sold out by this time, and before long Abbott did also. Each ranch better managed its own newly fenced portion and in time began to regain the good grass cover of earlier years.

Well ahead of the 1 July holiday each year, the Hamilton emptied until the next spring. Coutlee's cowboys moved Guichon's and Douglas Lake's cattle—the bulls joined them on 30 June—in manageable drives of 250 or less through the Coal Field until they met up with the Guichon crew, who took the cattle over the Courtenay Lake crossing. Then another Douglas Lake crew would take charge of the cattle at the Portland Ranch cow camp in Aspen Grove and scatter them onto the summer range. Douglas Lake and Guichon Ranch also shared summer range near the little town of Brookmere and around Kingsvale, country on the northern slopes of the mountains separating the Nicola Valley from Hope in the Fraser Valley. It was the task of the joint Douglas Lake-Guichon crew camped at Aspen Grove to ride these cattle for late calvers, to keep the bulls scattered and the range salted, and to pull out any cattle stuck in mud holes. Following the division of the Hamilton in 1940, Guichon Ranch decided it could do better running its cattle separately the whole year, and the joint camp that had been so beneficial for roughly two decades split up, as did the various Aspen Grove summer ranges.

With months of working seven days a week behind him, Joe Coutlee would rise one morning in midsummer, the slackest time of year for his crew, and roar, "Today's Sunday." It might or might not be Sunday, but everyone took the day off.

Of all the stories, tall or true, that have come down over the years, most relate to Coutlee's small summer crew during their drinking sprees. Once Annie Logan, one of the cooks, overheard Coutlee discussing her availability to the cowboys; she promptly earned the title One Punch Annie by knocking him down with a powerful blow to the chin.

Through the summer, some laid-off riders turned their hands to haying. The small shipping crew rode steadily, however. Marketable cattle have to travel slowly and gently to avoid getting excited and losing any of their grass-fattened weight, so it was a job for someone who was not in a rush. J.B. Greaves's son Joe, who had seldom worked at Douglas Lake while his father was there, filled the bill admirably as shipping boss.

On slack days, some of the cowboys would try to ride the rough horses, roping them by the neck and throwing them down in order to saddle them before climbing aboard. And if things were not lively enough, Joe Greaves or Joe Coutlee, Jr. would spur a horse in the flank when its rider was not looking and set the horse bucking and twisting in the air with its rider trying desperately to stay in the saddle. Young Joe Greaves was even known to whitewash a bay horse to fool its owner.

The Quilchena Hotel had the bar closest to the Nicola station; many of the valley's ranchers frequented it until prohibition closed its doors in 1917. Some of the liveliest evenings ever enjoyed in the Quilchena Bar—which boasted brass footrails, spittoons and three authentic bullet holes—followed the 1 July race days held on the course adjoining the hotel. Joseph Guichon's second son and namesake was the hotelier, and his own bouncer and lawmaker. Something of a dictator, Joe's method of dealing with an overexuberant customer was to lock him up in the root cellar behind the hotel. An hour there quietened the most belligerent drinker.

Even when prohibition came in, Joe Coutlee would stop off at the hotel after shipping to have a drink from Joe Guichon's private liquor stock, for Ward insisted that his men could drink only off the ranch. But one or two bottles always found their way into the cow camp, carried in the legs of an extra pair of pants knotted at the cuffs and thrown astride the horse. Ward had no effective way of stopping this practice other than to catch and fire the offenders and he was seldom at camp to do that. Moonshine came in too, which Coutlee did little to discourage. Once he accepted ten gallons in exchange for the loan of some of his JK horses. The distiller, Three-Fingered Slim, was a perfectionist and always sampled the batch to be sure it was right. One day he neglected to scour out the drum properly before using it—and he died. The drum had contained arsenic from the Grasshopper Control Campaign mixture.

To wind up these drinking sprees, Coutlee might roar, "Line up, you sons of bitches. You're not working for me, you're working for Douglas Lake Cattle Company, hereabouts a pretty big concern."

In June 1938, an epidemic suddenly broke out among the two-year-old beef steers in the Six Mile pasture south of the Hamilton Commonage, and 52 died. Coutlee immediately suspected that they had succumbed to larkspur poisoning, but federal and provincial veterinary surgeons diagnosed blackleg edema. Ward was convinced that the ranch had received an impotent batch of blackleg vaccine, and was prepared to sue the drug supplier. Then it transpired that Ward had personally picked up the vaccine and stored it in his car while playing in a weekend polo tournament; the very hot weather had probably deteriorated the vaccine. The drug supplier threatened Ward with a libel action. While these antics continued, Coutlee quietly moved the steers to the east end of the Wasley. This field was free of larkspur, and the deaths stopped, for the wild delphinium had indeed been the cause.

One November right after the first frosts had nipped the higher elevation grasses, as Coutlee was leaving Merritt for Courtenay Lake to help bring the herds home to the fall ranges, a Quilchena Indian friend demanded a drink from the sackful of bottles Coutlee had tied to his saddle. The illegality of giving liquor to Indians did not bother Coutlee, but he was in an ugly mood and refused. The infuriated man jumped up behind Coutlee on his saddle, pinioned his arms and stabbed him ten times in the back before running away. Dr. J.J. Gillis, the doctor in Merritt's hospital, had seen some incredible sights in that frontier town, but even he thought Coutlee's survival was a miracle. If the blade of the knife had been one inch longer, Coutlee could not have lived.

Once the cowboys had taken the cattle across the Merritt-Princeton highway, they drove them to the Minnie Lake country, for the Dry Farm crew to round up. They did the first weaning of 1,200 to 1,500 head in the Minnie Lake corrals, and allowed the calves to have their run of China Flat—an adjoining bunchgrass area. The weaned calves followed eight or nine faithful old lead cows, which Coutlee kept for just that purpose, down to the Home Ranch. Then the unweaned cows and calves, and the weaned cows came down in smaller groups to graze on the fall ranges around the Home Ranch and Chapperon.

One year an early winter snowstorm blew into a blizzard, which was unusual for October, and started 2,000 head of cows and their calves drifting determinedly towards the Home Ranch. It was a dark night, and even if it had been daytime Coutlee and his cowboys could not have seen farther than 50 feet ahead because of the slanting, driving snow. The riders were unable to turn the cattle which eventually reached the Home Ranch after taking out every single fence for miles. It was too late to return the cattle to Minnie Lake over the route they usually took in spring, so feeding started early that year and fencing continued late.

The daily responsibility for the cattle now moved from Coutlee's shoulders to the farmers', and the cowboys relaxed in order to enjoy Christmas. When Brian Chance became manager in 1940 he sent the crew to town for the festive season and reported to Frank Ward, "Old Joe looks very fit. He and all the other cowboys went to Merritt for Christmas, and treated the town to a first class exhibition of rough stuff, which was a great deal better than doing it here. It appears to have done them all a lot of good." During the same season a few years later he wrote, "Half the cowboys ended up in jail. Old Joe of course was rolled for a couple of hundred dollars as usual, after having drunk himself into a stupor. All arrived back after the New Year. Coutlee intended to lay off this winter on account of the foot which bothers him considerably in the cold weather, but as it has been so mild he has not gone yet."

Two years after, Coutlee thought again of laying off, because of a cancerous growth on his neck. Brian Chance recounted one of the last jobs he did at Douglas Lake, in November 1944. "Joe moved over from Courtenay Lake last week and is now cutting out for weaning. He tells me there will be about 450 of a fall brand, which is about right, as he turned out in the neighbourhood of 850 heavy and dry cows last spring." Joe Coutlee was just two weeks short of his seventy-seventh birthday.

Joe did not ride that winter—he was under the eye of Dr. Gillis in Merritt—and although he returned in January for a short time, it was the beginning of the end. Muggins, his wife, had died of pneumonia some years earlier. On 3 July, Joe Coutlee entered St. Paul's hospital in Vancouver. Three months later an old friend visiting him asked him when he was going back to the Interior. "Thursday," said Joe, and he did return on Thursday, but in a coffin. "The passing of no potentate could have produced more genuine grief," wrote the *Merritt Herald* after the funeral. Old Joe was buried in the Indian cemetery at Shulus, not far from his childhood home. People gathered from every ranch and every reserve for miles around to attend the last rites of the superb cowman whom they had been proud to know and whom Frank Ward in his eulogy called "The most valuable man that ever worked for The Douglas Lake Cattle Co. Ltd."

CHAPTER FIFTEEN

*Well! Brian, you now have the responsibility entirely on you of
running the Ranch. All good luck & good wishes for your success.
Don't give a damn for any criticism, also think 'big'. Don't let
trifles... or petty thoughts annoy you. Drive as hard a bargain as you
wish, but always let your word be as good as* Your Bond *no matter
what it costs.... As far as possible forget I.*

Francis B. Ward to Brian K. de P. Chance,
4 May 1940

Brian Chance grasped Douglas Lake's reins as World War II
broke out, and Canada's war effort profoundly affected his
first decade of management.

His first three winters were mild with unusually light
snowfalls that resulted in a water shortage on the ranges and
the hayfields. The animals took longer than usual to reach
market weight, but fortunately, a high demand for beef in
Canada, the United States and Britain kept prices rising. Pat
Burns and Company's bids for cattle were now in competition
with those made by other meat packers such as Swift
Canadian Company, and with those made by commission
agencies such as Weiller and Williams Limited, whose Ed-
monton partner, Lee Williams, would send many thousands
of Douglas Lake cattle to buyers across Canada and the
United States in the years to come.

Manpower was another matter altogether. The ranch
labour force absorbed much of the increasing profit as it
became harder to obtain and more expensive to keep. Chance
had forseen this labour shortage and had tried to prepare for
it. A larger than normal stock of ranch lumber had been cut
for maintaining fences, flumes and buildings and held out
well until labour became so scarce that Chance had to
temporarily abandon this repair work.

The letting out of hay contracts eased Douglas Lake's need

for crew, but even the contractors had difficulty finding men. In the summer of 1942, one of Chance's hay contractors lured Guichon's hay crew out of the Merritt beer parlour to Douglas Lake's hayfields by offering them a higher wage of $3.50 a day. Chance insisted that Guichon's men be returned.

There was a spate of government regulations and restrictions in late 1942. An embargo on exporting cattle to the United States deprived ranchers of much of their bargaining power and softened the market. A low ceiling placed on cattle prices came perilously close to the cost of production, which was continually rising. Rationing to two pounds of beef a week encouraged Pacific coast consumers to buy only the highest grading steers and heifers, and lower grades became harder to sell. Even the cow market in eastern Canada could not help much because of the high freight costs for western beef. An order prohibiting the hiring of labour from outside the province worsened the formidable labour problem.

As each new restriction was imposed, Douglas Lake's manager somehow coped. Brian Kesteven de Peyster Chance was a tall, lean Australian of good background, education and ability. When he had completed his schooling in England in 1921, Chance had returned to Australia, where his father owned three sheep stations, and en route he took advantage of his father's friendship with George Ward, Frank's brother, by visiting Douglas Lake for a few months to gain experience. Chance had then gone jackarooing (learning the business from the ground up) on various large sheep stations in Australia over the next five years before returning to Douglas Lake as understudy to the manager. When he took over, the deeded acreage covered 142,770 acres.

Despite the European war, the ranch continued exhibiting its grain-fed steers at the Fat Stock shows held in conjunction with the Vancouver Winter Fair and the Kamloops Bull Sale. Douglas Lake carried home many prize ribbons, but the greater prize came from the sales, which the War Measures price ceiling overlooked. In 1943 bidders at these shows were paying as high as $16.25 per hundredweight of live animals, so the competitors realized a substantial profit.

The war effort enhanced another sideline: hog production. Ranch labour built a large piggery at Chapperon to accommodate 90 brood sows, which farrowed twice a year. The weaned pigs fattened at Chapperon and at the Home Ranch on prairie grain; the federal government's temporary Freight Assistance Policy absorbed the cost of shipping grain out of the prairies for the purpose of finishing livestock. At six months of age, the market hogs went to Swift Canadian Company. In this way, Douglas Lake sent over 400 hog carcasses, valued at over $10,000, to Britain each year as its share of Canada's bacon quota to the armed forces. Another 150 hogs fed the ranch and satisfied neighbouring purchasers.

The increased program kept Lawrence Graham so busy that he questioned making bacon, ham and sausage for local consumption any longer. When he lost the ends of two of his fingers in the sausage machine—no one cared to eat that batch of sausages—the decision was made for him. The raising of hogs, however, continued.

On two occasions, slight traces of tuberculosis showed up in certain organs of the Douglas Lake hogs but Dominion health inspectors condemned only the suspect organs. After the condemnations, local vets carried out tests and every pig passed inspection. But when a second report still showed presence of the disease, Chance called in Dr. Wallace Gunn, provincial veterinarian for B.C. Dr. Gunn took one look at the Home Ranch yard and the hog fattening pens and said, "There's your trouble." He pointed to the hundreds upon hundreds of pigeons flying and nesting all over the Home Ranch, whose ancestors Willie Ward had released at Douglas Lake some years earlier. "You've got avian TB among your pigs. You're going to have to get rid of all those pigeons."

When Frank Ward and some of his old shooting buddies came up for the hunting season they spent a day shooting the pigeons. As Ernest Chamberlain remarked, "It was just like Dunkirk!" Four hundred pigeons fell to the guns that day, but the rest flew away when they at last associated danger with the noise of the guns. A second day's shooting brought down another 200 or so, then the blacksmith, Jack Campbell—an ardent hunter—finished them off single-handed.

Well aware that steady duck hunting on range lakes interfered with the cattle's watering habits and feeding patterns in fall, Chance had become an opponent of shooting, especially the occasional slaughter of huge flocks of wild ducks. He had first suggested in 1939 that the provincial Game Department turn Minnie Lake into a waterfowl sanctuary, which the game regulations for the 1941-42 hunting season authorized. Many thousands of ducks, geese, swans and pelicans flocked to the safe 2-mile lake once the hunting season began. Some keen hunters adopted the tactic of scaring the birds up off the sanctuary and over the guns waiting just outside the prescribed area. Game officers became aware of this infringement but, perhaps because of the difficulties in policing such areas, stalled for some years on enlarging the sanctuary as Chance wanted.

Having lost Frank Ward, the ranch's wild duck provider, and Minnie Lake, their fowl pantry, the farm crew at Minnie Lake soon tired of their monotonous beef diet. They asked their Chinese cook for some variety, and he obliged by carrying a stovepipe down to the shoreline of Minnie Lake, closing one end and dropping handfuls of grain into the other. He sat quietly in hiding until half a dozen ducks had pecked at the grain, each waddling into the stovepipe. The pipe fitted around them tightly, causing their legs to double up under them and preventing them from backing out. That evening, the crew cheered their ingenious cook as they forked up mouthfuls of tender roast duck.

In 1943, various government regulations, including the embargo on cattle export to the States, put the ranchers under the control of the packing house buyers. Brian Chance and others like him decided it was time for action; their provincial industry was worth $25 million and was annually marketing $5 million worth of livestock, so it should have been able to deal with the packing houses with which it did business. They obtained a mortgage from the provincial government for $25,000, bought the stockyards and other property of the B.C. Livestock Exchange at Vancouver, and began soliciting ranchers' support in the form of shares in the group they called B.C. Livestock Producers Co-operative Association.

Ranching shareholders hired a general manager and field staff, outlined a marketing policy and watched as subscriptions started pouring in. Douglas Lake purchased 50 of the $10 shares. A board of directors comprised of the same men on the board of directors to the B.C. Beef Cattle Growers' Association voted in Brian Chance as the first president of the Co-op, an office he held for 25 years.

Business began in October 1943 and the Co-op paid dividends within three years. By 31 March 1948, working profits had reduced the government mortgage by $19,000. Large and small ranchers had equal voices, and together they commanded power; the packers started to pay attention.

A heavy grasshopper plague deprived Nicola Valley cattle of much needed pasture in 1943. Because of the war, the Grasshopper Control Committee experienced difficulty in obtaining insecticides, in hiring the trucks to scatter the poison bait, and in hiring labour to man the trucks. In previous years, the *Camnula* (Roadside) species of grasshopper had been the most prevalent, and because the females laid their eggs in large egg beds, control was simplified. But in 1943, there were more of the *Melanoplus mexicanus* (Red Leg) species, which having no organized communal egg beds, presented a far greater threat.

Lawrence's son Laurie Graham, who was supervising the grasshopper control camp from the Dry Farm in 1943, recognized the cyclical nature of the grasshopper infestations and that next year the plague would be even worse. He ordered mechanical bait spreaders and prepared for a greater infestation. He was right, for one of the worst grasshopper outbreaks in the history of British Columbia hit the valley in 1944. Even with easier availability of insecticides, trucks and labour, the plague was out of control. Chance commented to Ward in July, "The grasshoppers are just about in command here now, and in places are shearing the range and in some cases the hay meadows, particularly the second crop, as effectively as a scythe."

At Courtenay Lake on a summer morning the cowboys would see a glittering mass of grasshoppers coming from Minnie Lake in the north. By the time the men returned from their day's ride, the land would be as bare as a table top. Trees took on an autumnal look as the pest stripped them of all their foliage. A layer of dead grasshoppers covered the surface of every well for miles. Horses refused to ride into the wind because of the grasshoppers that flew into their faces. When tents were taken down to move camp, the chewed-up ground sheets fell away. It was not safe to hang washing on a line, or to leave a saddle or ropes outside. The gravel roads were oily where car tires had squashed the grasshoppers to a pulp. On the railway line their crushed bodies greased the tracks, preventing trains from braking safely.

Each day men scouted the countryside, putting out the poison mix for the mature insects and checking the previous year's egg beds. Diesel oil and water put an end to each new batch of Roadside grasshoppers found squirming to life; the Red Legs went unhindered until mature. The campaign poisoned many millions of grasshoppers and saved many crops; still, the damage was tremendous. The total expenditure of the Nicola Grasshopper Control Committee was $17,200, of which Douglas Lake's share was $2,143.

Since Chance had become manager, he had tried—using haying contractors—to maintain annual hay crops between 7,000 and 8,000 tons, but a shortage of irrigation water followed by this grasshopper plague was keeping the annual crop down to 5,800 tons. Since Douglas Lake farmers were feeding out an average of 8,200 tons each winter, Chance had to buy the difference locally. He had to be careful how he bought, too, for once when buying a stack by volume he had paid for an empty wooden frame that enlarged the stack from the inside.

In 1944, Douglas Lake farmers cut and stacked another small crop of 5,772 tons. Six hundred acres seeded to barley and oats should have yielded 450 tons of grain, but only 100 tons matured. Deprived of the grass necessary to reach shipping weight, hundreds of beef cattle had to be placed in feedlots to finish on grain brought in from the prairies. Hay and grain purchases amounted to over $19,000.

In 1944, Chance became president of the B.C. Beef Cattle Growers' Association. Talbot Bond, foreman at Norfolk, retired in June after 23 years with the company and, after market gardening in Victoria for a few years, returned with his family to England. Chance moved Findlay Anderson of Bawlf, Alberta, the successor to Leith at the Home Ranch, up to Norfolk. Morris Roth, another farmer from Bawlf, took over from Anderson as foreman at the Home Ranch.

The country was alive with grasshoppers once more in 1945; the invasion was even worse than the previous year. Laurie Graham supervised from the Dry Farm, Raymond Graham ran a camp at Stump Lake, while Tommy O'Rourke ran a third control camp near Douglas Lake Home Ranch. Chance was forced to hold more cattle through the winter for grain fattening, and the 1,200 steers that he shipped off grass weighed 45 pounds lighter than normal. The pests had ruined practically all the grain, so the ranch had to purchase $15,000 worth.

As Douglas Lake suffered through the destruction wrought by grasshoppers, other matters distracted Chance. Ed Godley, the old harness maker, died. In the summer, Guichon's hay crew went on strike for $4.00 per day, though they settled for less; Douglas Lake quickly raised its haying wages. And then Joe Coutlee died in the fall, and every ranch employee and neighbour sorrowed.

Slim Dorin had been Coutlee's disciple for some years, so it was the most natural thing in the world when he swung into the cowboss job. Facing the fact that there could never be another Joe Coutlee, Chance learned to accept the ranch's loss and to appreciate Dorin's many capabilities.

The ranch was gradually procuring more farm machinery. In 1946, Laurie Graham opened the Home Ranch machine shop and became the ranch's first full-time mechanic, thereby lessening the load for the blacksmith and his forge. This move put Raymond Graham in charge of the grasshopper camp. The 1946 plague confirmed that the grasshoppers had reached their peak the year before. In addition to those killed by the control crews, many of the adults became highly parasitized so that they could neither fly nor lay eggs, and subsequently died. The next year, Raymond oversaw the first spraying of grasshoppers by airplane. The residual poisons used resulted in almost 100 per cent eradication of the remaining grass-hoppers. Modern chemicals ended at last the frightening nine-year cycle which had contributed to making ranching such a gamble.

But Douglas Lake lost some range in 1946. During the 1890s, the Duke of Portland had put together a ranch south of Forksdale. In 1912, Aspen Grove Land Company share-holders incorporated to acquire the Portland Ranch and further lands in the area totalling 15,500 acres. Since the '30s, Douglas Lake Cattle Company and Guichon Ranch had been renting the Aspen Grove land jointly for summer grazing.

Ed Mapson of Peat, Marwick and Mitchell, agent for the absentee Scottish owners of the Aspen Grove Land Com-pany, received an offer to purchase early in February 1946 and invited Douglas Lake to put in a bid. Chance offered $20,000, but Adrian S. Baillie, president of the Consolidated Granby Mining Company, had bid more and became the new owner. Matters did not end there.

Loss of this key summer grazing range at Aspen Grove meant also the loss of Douglas Lake's headquarters there. First Chance and then John V. Clyne, an experienced admi-ralty lawyer who was the new cotrustee and lawyer for Douglas Lake Cattle Company, tried to purchase an alternate site for the Aspen Grove camp. In doing so, Chance heard that Baillie was offering to buy up a number of other small operators. Fearing that Baillie might freeze Douglas Lake out of Aspen Grove altogether, Chance bought a quarter section owned by the Jesus Garcia estate and a half section owned by Frank Garcia. Both lots had buildings in useful locations, suitable for summer camps.

It seemed that Baillie was going in for ranching in a big way, for Lawrence Guichon and his brother Joe considered selling out to the mining president. Then it became known that Baillie was associated with Aspen Grove Lodges, Limited. This company proposed developing a vast hunting, fishing and skiing resort on the Crown land at Aspen Grove. Frank Ward rushed to Victoria, where he received assurances from the minister of lands that the government would not allow any such new venture to interfere with the established business of cattle raising.

In late March 1946, Baillie offered Chance and Guichon grazing rights on the Portland, and cattle from the two herds moved back to Aspen Grove.

Chance was just beginning to wonder why Baillie had bothered buying the Portland at all when loggers moved in: Baillie had bought the land for its timber value, not for its grazing. Chance and Guichon, who had never seriously considered the trees growing in such abundance on their ranges, were amazed.

Frank Ward, past 70 and pragmatic as ever, wrote to Chance the next year, "I fear our being only Stockmen, and thinking in such terms, we lost out badly on the Portland property. I met the man yesterday who told me . . . they took an option from Baillie for One hundred and twenty thousand dollars, which they expect to take up . . . . It seems to me we shall have to get a better idea of our own timber—what we own, and that adjacent that we don't own—and to what extent the grazing branch of the Forest Branch will seriously protect the grazing in their forests." Yet while the Ward family owned Douglas Lake, they never had the timber cruised nor made any attempt to sell it.

As always when one crisis passed, another was ready to take its place. In the summer of 1946, the labour problem worsened, but in a new way. With the war over, transients filled the countryside, drifters who seldom stayed more than a few weeks at a time. It seemed as though Douglas Lake had three crews—one coming, one working and one going.

Times were good and work that paid more than the low ranch wages was plentiful, so dependable men were at a premium. Occasionally, Chance would apply to the Kamloops labour office for half a dozen workers, but more often than not he would let them go in less than a week, having found them useless. He wondered why the labour office's selection process was so bad, for he had better success with men who just dropped in for work. Eventually he discovered that when asked for men, the labour office sent a truck to the police station to collect any troublemakers that needed to be run out of town, and thus a bunch of drunks, petty crooks and wasters would land at Douglas Lake's door. Chance stopped phoning the labour office for hay hands and managed as best he could with the labour that came and went.

With price ceilings now lifted from every commodity that the ranch had to buy, prices soared, but the beef ceiling remained in force. In early July 1947, Chance stopped selling cattle because he had to get prices that were too high for the packers to pay; they could not sell under the ceiling without losing money. In August the ceiling came off.

The year 1947 marked the beginning of fall and winter air searches for cattle which the cowboys had missed. The bush pilots who flew these little range-flying planes were an adventurous lot and never turned down a little excitement, as Brian Chance discovered on the day he and a pilot named Taylor touched down on frozen Ellen Lake to visit Harry Chapman at nearby Hatheume. On leaving, not only did Chance have to rock the plane's skis loose and then scramble onto the moving plane, but he also had the hair-raising experience of seeing the 65-h.p. Aronka barely clear the lodgepole pines in the thin sub-zero air.

When Chance went away on business in 1948, Frank Ward came up to Douglas Lake to keep an eye on the operation. He was loath to do so, however, because he had suffered a slight heart attack. His declining health, his 74 years, and the worry of someday having to replace Brian Chance made him think once more of selling the ranch; Douglas Lake had been off the market during the war.

Ward asked Chance's opinion on selling the ranch. "There is considerable activity among American buyers of ranch property in the Province just now," wrote Chance in July 1948, "which may be influenced in part by the prospect of the lifting of the embargo on cattle shipments to the U.S.A. likely to take place this Fall." Once the contract with Great Britain had terminated, Canada found herself with a surplus of beef, which the removal of the American export embargo alone could relieve. This did not happen until August 1949, but in the meantime three big Cariboo ranches, including the famous Gang Ranch, sold to Americans.

Chance's letter went on: "The property here has never looked better. The hay crop is heavy. One could mow grass on the ranges anywhere, and there are no grasshoppers. We have branded 3,000 calves already this spring, and it is likely there will be in the neighbourhood of 350 more this Fall. Ranch buildings are in good shape and up to date.... " Indeed, since 1939, ranch lumber had built 15 new buildings: seven foremen's residences, new cookhouses for Chapperon and Norfolk, a bunkhouse at Harry's Crossing, a new slaughter house, and four other structures. Wrote Chance,

Fences [are] in a good state of repair, and we are equipped with such modern machinery as is consistent with economical practice. Under these circumstances I think I could show anyone a million and a half dollars' worth at the present time....

From a sentimental point of view ... it would be a pity to see the place change hands, probably because my associations with yourself and the Company have been such happy ones, but one cannot mix sentiment with a million and a half dollars, which I would advocate accepting if it is to be had. The only advice I can offer is not to accept very much less.

Since 1942, unhappy with the pressures of her husband's job and the hold the ranch had upon him, Audrey Chance had been living in an apartment in Vancouver, only staying with Brian at the ranch during school holidays to see their son, Guy. At the end of March 1949 they divorced, Audrey returning to her birthplace, Australia, Brian and Guy staying at the ranch.

Brian Chance remarried that summer. His second wife, Jean Hay, petite and vivacious, took to the life and its pressures with great enthusiasm and panache. She hired Indian girls from Spahomin, dressed them in maid's white and gradually taught them how to run a household as they helped in hers. One who appreciated the invaluable training, Monica Tom, later certified as a hairdresser and attended happily to a stream of customers. She married Jacob Coutlee, a great-nephew of old Joe and an excellent cowboy, who had Joe's innate knowledge of stock and who could name and recite the pedigree of every horse in the Douglas Lake remuda. Jake was to become one of Douglas Lake's best cowhands in the '60s and '70s.

In 1949, when Lawrence Graham was beginning to feel the strain of running Chapperon and Harry's Crossing, he put his son Raymond in charge at Chapperon, and, later, his son Ralph at Harry's Crossing. Raymond Graham was just 24, newly married and something of a hothead. He had inherited his father's quiet ways, however, so that as he gained experience and age, he became an excellent foreman for Chapperon, fair to his men and contagiously enthusiastic.

Farming methods were beginning to change. Tractors experimentally replaced the two-horse teams on a bull rake, but in 1949, five New Holland balers purchased for a total cost of $14,000 did away with the need for hay contractors altogether. Two-horse teams continued to do all the stacking, however.

Slim Dorin, Douglas Lake's third cowboss, left on New Year's Eve 1949 to take the cowboss job at Nicola Stock Farm, which was closer to town. He was a good range cattleman and had run the cow camp economically. Three years later he became "All Around Cowboy" at B.C.'s rodeo finals.

Once more, Chance needed to find a new cowboss, and the obvious choice was Eddie Merino, the cowman who had been born in the Morton. Eddie had come back to the cow camp from the prairies in 1948 after a 46-year absence from Douglas Lake. A list of the ranches where he had worked in that time took in the best of Alberta's and Saskatchewan's spreads—76 Ranch at Crane Lake, Rod Macleay's Rocking P Ranch, the Gilchrist Brothers' $\overline{\text{X}}$ Ranch, Archie MacLean's CY Ranch, the Maunsell Brothers' spread at Fort Macleod, George Lane's Bar U, Alec Gillespie's Bar N Ranch, Gordon, Ironsides and Fares, and the Maple Creek Cattle Company. He had worked at the Dominion Range Experimental Station east of Manyberries and served in World War I.

Shorty was his nickname in Alberta, for Eddie stood no higher than 5 feet in his riding boots. In winter, loaded down with woollens and mackinaws, chaps and overshoes, he would throw his rope over the saddle horn and work his way up until he could reach his foot into the stirrup and climb aboard. Once on his horse, he was there to stay.

Eddie's position as cowboss at Douglas Lake was temporary from the start. It was too big a task for a man aged 66 to take in hand properly when he had been so long away, so Chance kept his eye on a young rider, Mike Ferguson, and wondered if he would be willing to shoulder the responsibility.

Eddie's first winter as cowboss, starting in January 1950, was long and cold. The drive of cattle going to Westwold, to be looked after by Tottie's son Robert Clemitson, began in pleasant balmy weather which changed drastically as the miles progressed. It was minus 52 degrees Fahrenheit when the drive reached its destination. Every member of the Douglas Lake crew peered through two slits in a gunny sack pulled over his head to stop his face from freezing; every member, that is, except Eddie Merino, who never went anywhere in winter without his all-enveloping buffalo coat. Robert's hired man, a German whom he called Deedee because Dietrich Heinrich Georg Schmietenknop was too much of a mouthful, stayed with the cattle on the feedgrounds until the cold spell let up, constantly opening new water holes which froze over again almost as fast as he cut them.

January's average temperature at Douglas Lake of minus 27 degrees Fahrenheit had the effect of drying up those cows with calves still at foot, so the cowboys kept weaning and the farmers kept feeding. Chance brought in extra hay from Westwold, over and above what the heifers down there could consume. In early April, Eddie turned out 2,700 head, but not many days after, more cold weather and 8 inches of snow forced him to bring them back to the feedgrounds. By May he and his cowboys had turned out all the cattle and by mid-June had branded 2,000 calves.

Chance dropped a line to Ward mentioning Eddie's progress and also inquiring about Colonel Victor Spencer, who had indicated more than a casual interest in Douglas Lake Ranch when he asked to see the company's balance sheet back in March. "Nothing more has been said about his proposed trip up here to drive around the property, but with everyone so busy it is probable he may lay this over until a little later on; that is if he is still interested. Have you anything more from Clyne?" Colonel Spencer had made known his intention to go to Ottawa to call on Jack Clyne, who was there to conduct the affairs of the Maritime Commission. The 67-year-old industrialist and philanthropist had been looking for new investments for his capital since 1948.

Victor Spencer had been born in Victoria in 1882, nine years after his father, David, began the dry goods business that eventually grew into a nine-store chain. The Boer War ended his formal education in 1900, and on his return he joined his father's firm, just a few years before it expanded to Vancouver. World War I took him overseas once more, fighting in several major French campaigns and earning him the designation colonel.

In the early '20s soon after his marriage to Gertrude Winch, Spencer acquired his father-in-law's Earlscourt Ranch at Lytton and there began raising prize-winning Hereford bulls. In between improving his position within David Spencer Ltd., encouraging the start of B.C.'s honey industry, investing in Pioneer Gold Mine, and advocating the building of the PGE railway, Spencer acquired the historic Pavilion Ranch, and the Diamond S at Dog Creek.

At one stage the Spencer Stores firm needed a large infusion of cash quickly. Having made a lot of money with his shares in Pioneer Gold Mine (which later combined with Bralorne), Colonel Spencer provided the funds. Thus when David Spencer Ltd. sold to T. Eaton Co. of Canada Ltd. in 1948, Colonel Spencer, director and vice-president of his father's concern, received the lion's share of the purchase price, reputed to run to eight figures.

Not long after Chance's letter to Ward inquiring about Spencer's interest in the ranch, Spencer called on Jack Clyne to discuss a purchase; then back in B.C. he met with the American W.P. (Bill) Studdert and proposed a deal with him over a bottle.

Studdert had a diverse background himself. He had apparently started work early as a deck hand on a fish boat; soon he owned the boat, then the fleet and then the cannery. In 1950, he had two freighters plying the west coast between the States and South America. At one time he made a great success of merchandising scrap metal and junk, and was involved in the livestock feed business. At Philipsburg, Montana, he owned the T Bar 3 Ranch, which ran 1,000 head of purebred Herefords.

In 1948, he and Floyd Skelton, an auction ring and stockyard owner of Idaho Falls, Montana, had bought the Gang Ranch for a reputed $750,000. There, Studdert gained a poor reputation among the cattle-raising community because of his roughneck tactics, starving of cattle, halving of grub orders, and stalling in paying wages after laid-off hands had made their own way down to his newly located office 85 miles away in Ashcroft. Although Spencer must have known Studdert's reputation, for they owned almost neighbouring spreads in the Cariboo, he chose to ignore it.

Spencer and Studdert looked at the available ranch statistics. During the Ward family ownership, Douglas Lake's land base had grown from 100,000 to 143,250 acres. This deeded nucleus controlled further government grazing lands capable of sustaining the 10,000 cattle and 500 horses on the ranch. The annual calf crop, averaging 3,050, allowed more than 2,500 head of mature stock to be sold each year. Cattle prices were rising fast, so that the 2,546 head sold during the fiscal year ending April 1949 had averaged the highest ever at $192 each, bringing in $499,539. This figure had more than covered the year's operating costs, even though expenses were rising as quickly as income: the monthly wage for a general ranch hand, for example, had just gone from $78 to $90. The year's net profit of $130,715 also included sales of hogs, hides and store goods.

The two men agreed to buy Douglas Lake in the early summer of 1950. Jack Clyne began rounding up the 400 ranch shares scattered between Vancouver Island and Great Britain. Only then did Spencer and Studdert spend a few days visiting

Chapmans' Hatheume Ranch, winter 1921-22 (courtesy of Rex Chapman)

Returning from cattle drive to Brookmere, 1928 (courtesy of Brian K. de P. Chance)

Johnny Guichon and Joe Coutlee rolling their blankets at Portland Ranch
(courtesy of Brian K. de P. Chance)

Crew at Hamilton Corrals (courtesy of Brian K. de P. Chance)

Mike Ferguson
(photograph by
Murphy Shewchuk)

Joe Coutlee and Joe Greaves fixing saddle in Raspberry bunkhouse (courtesy of
Public Archives of Canada, National Film Board Collection)

Courtenay Lake cookhouse when Slim Dorin was Aspen Grove foreman, circa 1943 (courtesy of *Toronto Star)*

Slim Dorin heeling in Hamilton Corrals, 1942 (courtesy of National Film Board)

Leaving the Home Ranch in spring (courtesy of the *Toronto Star)*

Brian Chance at Douglas Lake and flying over the property in Studdert's plane. Actually, Spencer was no stranger to the ranch, having hunted and fished there at different times in the past.

Satisfied that everything was in order, Spencer and Studdert purchased the shares in Douglas Lake Cattle Company from the Ward family in late August for $1.4 million.

Had Cecil Ward still been alive, he might have protested the sale, considering that his father had turned down the 1920 offer of $1.5 million and that the ranch had since bought another 22,250 acres. But the purchasers were more interested in their rate of return, which, using the figures for 1949, would have been a 10.7 per cent profit on a $1.4 million investment. The new owners looked forward to such profits continuing.

Though Frank Ward agreed to remain chairman of the board, his part in running Douglas Lake ceased with the sale. The man who had treated Brian Chance as a son died at home in Victoria on 27 April 1953, aged 78.

CHAPTER SIXTEEN

The high spots.... The stirrup high grass that could hide a half grown calf.... A sense of 'Empire'. One could ride a day or two in almost any direction and then might not find a neighbour.

Perhaps above all, the Wilderness. I suspect in the beginning we were too naive to be intimidated by it, and so came to love it.

And I still do.

Notes on Hatheume Lake by Rex Chapman,
1973

The main valley of the Nicola Lake and River system lies northeast-southwest. Parallel to and southeast of this area lies the interrupted rectangle of open bunchgrass deeded to Douglas Lake Ranch. Farther southeast of that rectangle, a crescent of timbered country stretches from Aspen Grove in the southwest around to Salmon River in the north. The height of land separating the Nicola Valley watershed area from the Okanagan Valley marks the eastern boundary of these wooded mountains that are potential summer grazing for Douglas Lake's cattle.

Greaves first used the north and south extremes—south of Salmon River and around Aspen Grove—as summer range. Caught up in the crescent between were numerous family-sized cattle operations whose wild meadow preemptions controlled the summer grazing adjacent. As these ranches came on the market, Douglas Lake purchased them: Smoky Chisholm's 320 acres south of Courtenay Lake by 1913; the Raspberry brothers' 800 acres around Pennask Lake, along with the rest of their ranch, in 1916; Wayne Sellers's 1,600 acres around Paradise and Pennask lakes in 1936; A.W. Wright's 200 acres in Pothole, southeast of Courtenay Lake, in 1939; Amil Peterson's quarter section at Tea Lake in 1940; Jerry Mellin's 456 acres south of Bull Canyon in 1940, and the Garcia land at Aspen Grove in 1946.

By 1950, only a few parcels of alienated land lay within this summer grazing range. Some, such as the ranches of Vic Shropshire, Scotty McLeod and Les Bryant, straddled the Merritt-Princeton highway. East of Aspen Grove was a cluster of quarter sections first preempted by Nels Wedin, Henry Shrimpton, Hans Halferdahl and others, and afterwards brought together into one ranch by Harry Gilroy and Ethel Stephens. Their 1,600 acres controlled 20,000 acres of grazing.

The only other alienated land in the crescent-shaped mountain country that spanned 60 miles was the Chapman family's 2,000-acre Hatheume Ranch, which controlled roughly 40,000 acres of grazing. They had put the property together many years earlier.

The formidable country around Hatheume Lake, 14 miles southeast of Douglas Lake Home Ranch and 4,500 feet above sea level, allowed only the toughest of homesteaders to work it. Several waves of would-be settlers came to this jack pine jungle between 1910 and 1920, attracted by the beaver-dammed peat meadows.

John Chapman, Freeman of the City of London and fresh from his London law practice, brought his wife and four children to Mazama in the Okanagan Valley in 1912. For his tiny wife, the change was immense from running an orderly English house and family with the help of four or five servants, to living in a log cabin, although the building of the Kettle Valley Railroad line from Princeton to Summerland via Mazama in 1914 may have appeared to her as a lifeline to civilization. John Chapman's first quarter-section preemption eventually grew into a 500-head operation that had Mazama as the home ranch, summer range around nearby Trout Creek, and wild hay meadows growing winter feed at Hatheume.

Gordon Sellers first sparked the Chapmans' interest in the high country. Finding himself "grasshoppered" out of hay in Princeton by 1913, Sellers and some of his neighbours looked to Hatheume's wild meadows. The "rip-gut" hay tided them over, and Sellers preempted land three miles south of Hatheume Lake the next year. In helping Sellers move his cattle to this preemption at the onset of winter in 1916, John

Chapman's son Rex first viewed the snowy winter reaches that constitute the height of land between the Nicola and Okanagan valleys. It seemed madness to be moving the herd up to the timber where the snows came earlier, fell deeper and stayed longer than at the lower elevations, but Sellers had to use the land as a winter feedground because it was too difficult to carry away the tonnage of wild hay that he had stacked there.

In 1918, Rex and his father preempted two wild meadows close to Hatheume, sight unseen. The two youngest Chapmans, Rex and Don, hiked in that fall to see these preemptions. They found a profusion of lakes, and their approach flushed into the sky a black chorus of migrating northern ducks. They did not see another living soul during the week-long trip.

Not until three years later did the Chapmans go back to cut their wild hay and stack it for use. The family now owned almost a square mile of scattered lands, surrounding and surrounded by longer-established settlers of whom November Gottfriedson was the oldest.

November and his brother, Henry, had ridden in from Douglas Lake in February 1910, and found that by a trick of weather the meadows of Hatheume were almost bare of snow. They had been searching for just such a place since leaving Idaho. Their father had been a Danish fisherman on the Fraser River, their mother an Indian.

Henry did not stay long at Hatheume, but November, and November's woman, Ellen Yeakey, remained to preempt a quarter section in the centre of a triangular space bounded by the Indian-named Hatheume Lake in the west and two smaller lakes in the east. The southeast one became known as Ellen Lake; Nellie's Lake to the Indians. Farther southeast, a 6,224-foot peak became Gottfriedson Mountain.

As the summer progressed, November, described by Rex Chapman as "a short stocky man built as solid and tough as a fir stump," widened the Indian trail from Spahomin. He and Ellen brought in a wagonload of winter supplies. Their first winter at Hatheume was long, and the accumulated snows looked capable of outlasting their groceries. They set off for

Quilchena, taking three days to cut a trail with a hay knife through the wind-encrusted snow covering their meadow. They had to feed the horses their mattresses of hay. When they reached the open range, they found that the bunchgrass had already grown 8 inches.

Pete Peterson, after searching throughout B.C., eventually found his Shangri-la in 1913 on the southwest shore of Hatheume Lake. A little coyote trapping, fishing and selling of smoke-cured eight- and nine-pound Hatheume trout; the occasional spell of cattle feeding at Douglas Lake and Guichons', and a small still in the back shed met "Bronco Pete's" few wants. For getting around, he used an old Nicola stagecoach and two incredibly small sorrel ponies.

He once confided to Rex that the Hatheume country was too tough for him. "I should have stopped at Ootsa Lake," he said, referring to a small settlement 350 miles northwest in central B.C. "It has everything a man could desire. A grand country." "Why didn't you stay there?" young Chapman asked. "I was young at that time, I didn't want to settle down." He had been 62 at Ootsa Lake.

The number of homesteaders increased in 1914 with the arrival of Harry Clayton, who did not stay long and became a blacksmith for various Nicola Valley ranches; Boyd Almon, a Nova Scotian who ran a trapline, took contracts on the local ranches in summer and seemed to be under a compulsion to repay any favour tenfold; and Phil Cameron, who preempted land on the south end of Ellen Lake. One day the Chapmans saw him nursing his jaw; all his front teeth were missing, and Gottfriedson's knuckles were red and sore. Not long after, Phil moved Ellen Yeakey in as Mrs. Cameron, and hung a framed marriage licence prominently on the wall.

Christopher Columbus Warmot, who built a cabin on the north end of Ellen Lake, never attained the navigational prowess of his namesake. Often eager to get home after a trip, he would try to find a short cut. Once he found himself stranded on a beaver house with no way to get off until morning revealed the trail. Another time when night overtook him, he decided to camp where he was, building a warm fire for comfort. The night was long. As the dawn light streaked the sky, Warmot peered through the morning mists to see the outline of his cabin just ahead.

The Chapmans began buying out some of their neighbours in 1921, starting with Phil and Ellen Cameron, whose log cabin sported two good-sized rooms. This cabin became the Chapmans' Hautheume ranch house. The buildings on other newly purchased lots were adapted to new uses, but they never did anything with one particular low cabin. A tall Okanagan inhabitant had built it and, finding his cabin shorter than himself, had dug down 2 feet into the dirt floor. This was fine until spring came with its accompanying thaw and rush of water. He apparently solved the problem by building a raft to keep himself and his belongings dry.

Bronco Pete died in the winter of 1921-22; he had taken a coyote capsule of strychnine to cure a toothache. The Chapmans bought his preemption, the sorrels and stage, and some traps. Their purchase agreement covered the still, too, but by the time they took over the property, a needier soul had taken it.

The Chapmans' occupation of Hatheume was only seasonal. The length of winter at ranch headquarters in Mazama determined the length of winter feeding at Hatheume. As soon as the fall ranges were white with snow, the Chapman herd trailed north. Hatheume was a lonely, quiet place in winter. The black bears were hibernating, the deer had moved down to lower elevations, the wild ducks and geese had flown to warmer climes, and the only companions left to the solitary cattle feeder—one of Chapman's three sons—were the cattle and his saddle and pack horses. A great rapport grew up between the feeder and his four-legged friends.

Each year some Douglas Lake stock would wander from their neighbouring ranges southeast towards Hatheume. Even though the Douglas Lake cowboys would ride in to collect such strays, there were some winters when the Chapmans fed 60 extra mouths—Douglas Lake cattle that had wandered in too late for the cowboys to pick them up. In the mid-'30s, Frank Ward and Don Chapman agreed on the location of a drift fence between the two outfits' ranges.

When the going became too tough for the saddle horses, the

Chapman brothers resorted to snowshoeing, at which they became adept. One year when Dave Lindley and the boys went to gather strays, the Hatheume snow lay particularly deep. The going was too hard for the horses, so Don Chapman loaned each cowboy a pair of snowshoes. They searched out the strays and drove them home, but they had great difficulty in staying right side up on their snowshoes. In the bunkhouse at Douglas Lake the incident, told and retold, gave birth to a legend: the Chapman brothers were the fastest men around on snowshoes; they could travel an incredibly tough six miles from Hatheume to Mellin Lake in less than an hour.

As the snows melted from the Chapmans' early spring range at Trout Creek, their herd was driven west to Minnie Lake, southwest to Aspen Grove, south along the highway to Princeton, and northeast to Mazama; only the most imminent calvers were kept at Hatheume into the spring. It was a circle tour, four sides of a pentagon, and over four times as long as the straight distance between Hatheume and Mazama. The cattle lost so much weight on the 100-mile drive home that it took them all summer to put back what they had lost. To avoid such severe weight loss, the Chapmans began using an old Hatheume-Mazama trail that had been blazed by Gordon Sellers, even though it was narrow, rocky and devious.

In 1925, the Chapmans bought out Gordon Sellers and November Gottfriedson, increasing their Hatheume holdings to almost 2,000 acres. Boyd Almon and his winter trapline alone remained.

One piece of land they acquired—a mile and a half west of Hatheume Lake—was close to the shores of Pennask Lake, a body of water teeming with fish. The Indians made two treks there each year, one at spawning time, when they set fish traps at the inlet to Pennask, and the other in winter, when they fished through the ice. With this acquisition came a right to water cattle at a place on Pennask Lake that soon became known as Chapman Bay.

Not long after, Dominion Fisheries first recognized the wealth within the waters of Pennask: wealth in the form of trout eggs that they could strip from the spawning females and raise for stocking other B.C. lakes. The Pennask Lake fish hatchery was built in 1928, the Chapman cowboys and government egg collectors finding themselves compatible. But that same year, another group discovered the wealth of Pennask Lake, and its ideas were not compatible with the watering of cattle.

James D. Dole of Boston, Massachusetts, the Harvard graduate who became the pineapple king of Hawaii, had been fishing in British Columbia eight years when in 1927 his favourite lake—Fish Lake—became fished out. The Cowans, who owned Fish Lake, began searching for another fishing haven at Dole's instigation. The steady supply of small rainbow trout at Pennask attracted them. Each spring the Indians took away over 20,000 fish, and yet they remained small, a certain sign that the fishing pressures would have to be excessive indeed to clean out the lake. James Dole confirmed this choice in the fall of 1927 when he visited. He also appreciated the bird-hunting and trail-riding opportunities that the area offered. He decided to create a private fishing club—life membership, $1,000—to supply and maintain consistently good fishing for any other Nimrods able to afford the luxury.

So in 1928, the Cowans and others began purchasing every foot of the 20-mile shoreline, as well as each of the 15 to 20 islands within Pennask's perimeter. Dole also purchased two of Gordon Sellers's nearby lots. He commenced the construction of a fishing lodge, wharf and breakwater, and widened and levelled the trail through Douglas Lake for a roadbed. Dole began soliciting his friends to share in his $70,000 expenditure, but the collapse of the American stock market severely hindered his progress at first.

Dole and his associates neglected to secure the shoreline of Chapman Bay. Fearing Dole might close off their watering hole to them, the Chapmans applied to purchase the 40 acres between the bay and their own lot. Dole's group retaliated by making a similar application. Antagonism grew and a deadlock ensued. At the resultant hearing, Frederick J. Fulton, Kamloops lawyer, former member of Parliament and Cabinet, and cotrustee with F.B. Ward for Douglas Lake, spoke for Pennask Lake Club; John Chapman spoke for his family; the Indians, through an interpreter, spoke for themselves. The hearing resulted in the Crown forever reserving the land from alienation, and for the use of all.

Their needs thus reassured, Pennask Lake Club, the Chapmans, the Indians and Dominion Fisheries cohabited peacefully. The provincial Department of Recreation and Conservation eventually took over the Pennask Lake fish hatchery operation, which in the '70s provided between 25 and 40 per cent of the annual trout egg requirements for the province's fish culture program. Pennask Lake Club remained a private fishing club, and the lake, allowing fly fishing only, achieved fame as one of B.C.'s best. But for the Chapmans in 1929, it remained a well-situated watering hole.

The original Cameron cabin burned down in 1930 but the $750 insurance came in handy, since steers were then selling for the pitiful sum of 3 cents a pound on the hoof. After the Chapmans had built a new log cabin, with the help of their neighbour Jerry Mellin, they were mortified when they understood the value of a small pen-and-ink sketch which had been nailed to the wall of the original cabin with a 4-inch spike. It had been drawn by the now famous Charles M. Russell.

Earlier in the year, before the cabin burned, and before haying, the Chapmans had found time to straighten the Hatheume-to-Mazama trail. Keeping to easier grades, taking out corners, avoiding the open areas subject to drifting, they took six miles off its original length. Finished, it was greatly improved both as a stock and pack trail. Only occasional trips via Quilchena were now necessary.

The task of opening the trail from Hatheume to Mazama in winter was now easier, but still a job that the feeder had to do after every snowstorm, every two weeks on average. An exceptionally heavy storm dumped 5 feet of snow at Hatheume one day in February 1935. The day-long job of trail-breaking extended to three as Rex and two pack horses laboriously packed down the excess snow on the Chapman herd's homeward route.

John Chapman made his last trip to Hatheume in 1930. His wife never went there. In 1931, eldest son Harry married and went into the house construction business, leaving Don and Rex to take care of the Hatheume operation; and quite often Don was away, packing for a surveying outfit or taking a haying contract on one of the neighbouring ranches. Thus, in 1936, when the enterprise was at its peak size of 500 head, the Chapmans promoted Reid Mueller to foreman.

As Rex Chapman neared 50, he had warnings of ill health and decided to listen. Since 1942, the Chapmans had been ready to sell their ranching operation. In 1948, the senior Chapmans moved to nearby Summerland. Gradually, they sold off their herd; yearlings one fall, cows and calves the succeeding year. They decided to sell the Mazama and Hatheume holdings in two parcels. Douglas Lake first turned down the purchase for fear of becoming "land poor," as Ward termed it, but when Spencer and Studdert took over, the brothers approached them. Being part of the Upper Nicola River watershed and fitting so naturally into Douglas Lake's southeastern boundary, Hatheume Ranch's 2,000 acres was a sound acquisition and so it became part of Douglas Lake Cattle Company in January 1951.

Although Hatheume was now part of Douglas Lake, it received minimal use; Rex retained his trapline in the area and soon prevailed upon Brian Chance to lease him a cabin site on Hatheume Lake. A Vernon man, Ray Redstone, had come to know the area when haying for Rex and had also hayed at Chapperon. With the complete approval of Brian Chance, Redstone set about developing a fishing resort at Hatheume. He fixed up the Chapman cabin, idle for seven years, to use as a temporary lodge for his few customers in 1958. Negotiating right-of-way easement across Douglas Lake's deeded land, he built roads to the piece of shore on Hatheume which he was in the process of obtaining from the Crown. In 1959, work began on Hatheume Lake Lodge, a superbly simple log building able to accommodate one family and 40 dinner guests. Rex came in handy with his logging experience, felling, peeling and skidding the enormous pine logs into place. Over the next seven years, while the fishing business multiplied and Redstone built more roads to the good fishing lakes in the area, Rex returned to help build six guest cabins, each nine logs high, a staff cabin and a fish house. Hatheume Lake became one of the finest fishing resorts in North America.

Douglas Lake Cattle Company was slow to open up and use the 40,000 acres of government grazing that its Hatheume acreage controlled, but by the mid-'60s its program was underway. By the '70s the Chapmans' old winter quarters, which had fed up to 500 head, became summer grazing for up to 1,600 head of Douglas Lake cattle. Douglas Lake had finally consolidated its ownership and tenure within the mountainous Crown grazing range, just as it had earlier consolidated its deeded holdings over the open bunchgrass plateaus. The ranch now controlled the entire Upper Nicola River watershed area, a crescent that stretched 60 miles in length and 25 miles in width.

Even as Rex, the sole surviving member of the Chapman family, neared his eightieth year, he continued to journey up to Hatheume. A pickup truck replaced his saddle and pack horses; Redstone's jeep road replaced the pack trail; a snowmobile replaced—though not entirely—his snowshoes; but nothing would ever replace what Hatheume continued to mean to Rex Chapman. Always it would be the wilderness land he had grown to love.

CHAPTER SEVENTEEN

There comes a time in the life of every Company when our astuteness as owners and managers is determined by the rapidity by which we can reduce expenses and this time is now here. I know you are alive to the situation but you have got to be drastic.

Frank MacKenzie Ross to Brian K. de P. Chance,
29 October 1953

When Colonel Victor Spencer and Bill Studdert bought Douglas Lake Cattle Company in August 1950, Spencer became president while Studdert became managing director and vice-president. They retained Brian Chance as manager.

Turmoil ensued for the next few months. Spencer, because he was a dyed-in-the-wool free-enterpriser, had told all his other ranch managers that if they could not sell his cattle without the assistance of the B.C. Livestock Producers Co-operative Association, he would find someone who could. The result was that Spencer and Studdert turned down all the benefits of participation in the Co-op. Their position was embarrassing for Chance who was still president of the association.

Eighty tons of cattle pellets, worth $12,000, arrived one day for Douglas Lake. Chance knew no reason for this wasteful expenditure until Studdert flew in and asked that the ranch pay for them: it seemed he had been unavoidably saddled with four carloads. Chance, his 1930s training evident, bristled as the ranch steers licked the pellets up like candy.

Studdert's flying trips to Douglas Lake, in his capacity as managing director, continued with increasing regularity, and in the fall, Frank Ross and his wife visited the ranch. Ross and Colonel Spencer were good friends and business associates, being codirectors of many an important enterprise, including the recently sold David Spencer Ltd.

When he left his native Scotland at age 19 in 1910, all that Frank MacKenzie Ross had to his name was his senior matriculation from the Royal Academy at Tain. His first job was junior clerk in a branch of the Canadian Bank of Commerce in Montreal. During World War I, Canada awarded Ross the Military Cross and Bar for outstanding service in the field. In 1919, he returned to the Canadian Bank of Commerce working in the head office in Toronto, but left very shortly after to work at Saint John, New Brunswick. There he helped form the Saint John Dry Dock and Shipbuilding Company Ltd. and by 1926 owned the firm, his first major business acquisition.

Success was a byword for Ross in business. The simple lad who would proudly claim years later that he had left Scotland relatively penniless, went on to make his first million in the '30s when so many were wondering where their next meal was coming from. Numerous directorships and board chairmanships of influential companies came his way.

Too old to go to the front during World War II, Ross took on a dollar a year job as director general of naval armaments and supply in Ottawa from 1941 to 1946. The wartime growth of the Royal Canadian Navy was unprecedented, and due in great part to the energy and vision that Ross brought to his position. The Right Honourable C.D. Howe as minister of munitions and supply, and Ross's direct boss, later praised the "sage advice, breadth of view and integrity" that Ross provided Canada's wartime effort. When war was over, he assisted in returning Canada's economy to a peacetime basis.

In 1945, Ross married Mrs. Phyllis Turner, widow of an Englishman and daughter of James and Mary Gregory of Rossland, British Columbia. That year the couple moved to Vancouver, Frank commuting to Ottawa to continue his work. In a rare husband-and-wife ceremony in 1948, Frank Ross received the Companion of the Order of St. Michael and St. George from the British government for his work during World War II; Phyllis Ross accepted the title Commander of the Order of the British Empire for her work as an administrator for the Wartime Prices and Trade Board. One of the two children from Mrs. Ross's first marriage, John Turner, years later became federal minister of justice and then minister of finance.

Towards the end of 1950, Ross, Spencer and Studdert met and agreed to split the shares in Douglas Lake three ways, but a short time later Ross changed his mind. He informed the Colonel that he did not wish to have any part in the enterprise if Bill Studdert were going to remain. Ross and Studdert's personality clash was not an uncommon occurrence among men who talked of a million dollars as glibly as a cowboy talks of ten. Since Spencer had brought Studdert in on the deal, it was his job to get him out.

The two original purchasers met in the Spencer offices at the top of the Marine Building in Vancouver to discuss the new arrangements. The meeting lasted for three days during which neither man left or had a change of clothes; only meals and liquor were sent in. Both Spencer and Studdert were suffering from complete fatigue by the time they reached a decision.

Studdert, having been a shareholder for a mere eight months, relinquished his third of Douglas Lake for a tax-free gain of $140,000; initial financing had been such, however, that he had never put up one cent. On 25 April 1951, the old company finally came to an end and two equal owners, Colonel Victor Spencer and Frank MacKenzie Ross, incorporated a new Douglas Lake Cattle Company Ltd. with an authorized capital of $2.1 million.

Even before the April incorporation, Spencer and Ross introduced changes. A new accounting system designed to provide more detail called for typing expertise which Chamberlain did not possess. As he had been bookkeeper at Douglas Lake already for 21 years, he decided to retire. Lyman Swennerton took over the new books.

At a meeting held on 20 December at Frank Ross's Vancouver home, Bill Studdert, who at that time had not yet left the company, suggested purchasing the Chapmans' Hatheume Lake Ranch. The 2,000 acres and improvements had been up for sale since 1942 for $15,000 but Studdert hoped $12,000 would be sufficient. It came into Douglas Lake's hands in 1951 at that price.

Frank Ross drew up a memorandum following this meeting which had as its core a method of financing the purchase of Douglas Lake. The proposal was that for the two-year period commencing December 1950, the ranch would sell 4,800 head of female stock over and above the normal annual sales of around 3,000 head of steers, heifers and cows. This would mean selling 50 per cent of the herd each year rather than the normal 25 per cent.

"The ranch will have had two crops of calves so that it is not considered that the suggested realization will be too disturbing," Ross explained to Chance. "Providing approximately the present price of cattle is maintained and the said number of cattle sold, the purchaser's debt will be in the main liquidated at the end of twenty four months."

The suggestion of selling so many cows initially horrified Chance until he decided that the new owners intended to cull the poorer cows from the herd thereby. But that was not their idea at all: they planned on getting $300 apiece for top cows in calf, in order to realize an ideal $1,440,000 and so cover their capital outlay for Douglas Lake. It seemed to Chance that the overall integrity of the Douglas Lake basic herd would never withstand Ross and Spencer's demands.

Such heavy selling meant that all 1,500 of the yearling heifers would have to deliver a calf as two-year-olds, and even then the calf crop would be much smaller than normal. Chance would have to open an extra cow camp just for the yearlings, and Eddie Merino, with all his experience, was the natural choice for taking charge of it. That left the job of cowboss unfilled once more. Chance thought of the various men on the cowboy crew and kept coming back to Mike Ferguson: he had the makings of a cowboss.

Mike had in him a little of his ancestor, Peter Skene Ogden, "the most energetic, the most dynamic, the most far-ranging...fur-trade apostle" to come to B.C. for the Hudson's Bay Company in the nineteenth century. But Mike also had the ability to arouse antagonisms and jealousies, and one day Jimmy Roubillard picked a fight with him. When it was over, Roubillard's face was in such a mess that the Merritt doctor had quite a job to stitch him up; Mike came away with few injuries. But the fight got him removed from camp, as his fellow riders, who hoped to improve their own chances of becoming cowboss, had predicted it would. Chance moved Mike down to the Home Ranch in June 1950 to be fence rider there, and to do the butchering. Still, in 1951, when Chance had to find a new cowboss, he gave Mike the job. The rather gaunt, black-haired, 32-year-old cowboy, one of the first in the camp to own a car, set out to prove his ability.

The yearling breeding program ran into difficulties in its first season. In December 1951, the 1,500 heifers brought down to Westwold were two- and three-year-olds. Although some calamity occurred every year, such as cattle freezing in their own tracks or snow drifting around the stackyard thereby allowing the cattle to help themselves to hay, this year all seemed to be going well. The three-year-olds were due to calve on 1 April, and well ahead of calving date the cowboys drove them home, leaving the 300 or so two-year-olds—due on 15 April—for another two weeks.

Since there had been a thaw, Robert Clemitson, who had taken over from Tottie in 1944, moved the Douglas Lake heifers into a drier, sloping meadow. Then he moved his own cattle to a similarly well-drained area. When he finally reached home, he was cold and tired, but the look on his wife's face told him that he would not be able to put his feet up for some time. "Douglas Lake's heifers... they're calving," announced Margaret Clemitson, her normally cheery face creased with worry.

Robert rode over and found his wife was right. Yet this was not a normal calving, for 150 heifers had aborted their calves. Heifers bawled and jostled and deserted their calves; some calves were dying, others dead; very few were up and sucking. One little calf that had toppled into a badger hole avoided certain starvation because Robert almost tripped over it. The mass confusion was too much for him and his helper Deedee to cope with, so he phoned Douglas Lake for assistance. Unfortunately, Brian Chance was away.

Meanwhile, Robert and Deedee rushed about, mothering up cattle, pouring all the milk they could get down the throats of those calves whose mothers either had little milk or no mothering instinct. They assisted in some births, and they tried grafting orphans on to milking heifers that had lost their calves. Robert had no time even to wonder why the heifers were aborting. Each day more calves came until there were around 300, all premature. Some of the heifers died, and Margaret became nursemaid to nine orphans.

Brian Chance rushed down to Westwold as soon as he had returned to the ranch. He and Robert toured the disaster area and discovered the source of the trouble. A lone ponderosa pine standing in the meadow a short time before had been felled by the feedground owner. Cattle go after anything green in the spring and the turpentine in the fresh pine needles scattered over the slope had caused the mass abortions.

Chance ordered five tons of grain to bring in the heifers' milk, and he sent down some feed troughs. Robert and his helpers managed to save most of the heifers and their calves, though 30 or 40 were dead before he could move the cattle into a field free of pine needles. It was 15 May before the cattle could safely return to Douglas Lake—the orphans were shipped in a pickup truck—and some cattle even stayed in Westwold all summer, returning only at the end of the next feeding season.

For four years the ranch bred yearlings, but Eddie Merino left before the fourth season was through. The losses—6 per cent of the heifers and 18 per cent of the calves—were more than anyone had expected, and the little Mexican-Indian found the work far too exacting. After a brief return to the prairies, Eddie became Gerard Guichon's cowboss. He died with his spurs on early in 1960, aged 76. Guichon arranged for Eddie Merino's burial in the Indian cemetery at Spahomin; a field rock engraved with the brands of all the prairie ranchers for whom he had worked was his headstone.

Because they had sold off the top end of the cows, Spencer and Ross had a herd which leaned heavily towards younger animals. In the fall of 1952, they kept up with the times by selling yearling steers as well as two-year-olds. These yearlings went to prairie feedlots for the grain finish that consumers were beginning to prefer to the grass-fattened finish. Thus in

1952 Douglas Lake again sold far more cattle than usual, with the result that in 1953 there were less than 200 head for sale. This coincided perfectly with the hoof-and-mouth outbreak on the prairies that year when the price of cattle all across Canada dropped a dime overnight, from 34 to 24 cents a pound.

The logging companies working on the Portland Ranch in Aspen Grove had completed their hauls of timber when Spencer and Ross began to deal with Adrian S. Baillie for the property. The transaction for the 15,500 acres of improved summer grazing finally went through in April 1953, for $50,000.

The previous year Spencer and Ross had purchased a half section, situated in the middle of their deeded land north of Pothole, from Archie Davis. Such purchases, which brought Douglas Lake's deeded land to 161,080 acres, continued the policy of consolidation that J.B. Greaves had advocated so many years earlier.

Right from the start, however, Ross had been putting the brake to other expenditures that Chance might make. It was not until 1953, though, when income potential was one fifteenth of normal, that Ross applied the brakes so hard that the ranch almost screeched.

In a February memorandum Chance agreed that he would not purchase any additional equipment or machinery, nor build any new fences or flumes. Money could be spent on maintenance only. In April, Ross judged the inventory held at the Home Ranch machine shop too large. He instructed Chance to cut it down to one month's supply, though the harassed manager was loath to comply, knowing full well how erratic was the supply of parts, and how a delay in obtaining one could delay work in the fields. Ross also halted the construction of any new building, shed or barn that April. Such restrictions stalled the normally slow but steady advancement of the ranch, until deterioration set in. As the state of range fences ran down, so did ranch morale. Even Chance's management job degenerated into a caretaker job.

The year's expenses decreased by a substantial $100,000, but Ross wrote again: "Still further reductions *must* be made this year in Gas and Oil, Repairs and Maintenance of Buildings and Equipment, and Legal and Audit Fees. It is hoped that the Wages account will be substantially reduced also." There was little more that Chance could do that he had not been in the habit of doing ever since he had become manager, as regards cutting expenses. The cost of almost everything from horseshoes to veterinary supplies was rising.

At the depth of this depression, the closest thing to a stampede that had ever happened at Douglas Lake, occurred. When Mike Ferguson took over as cowboss one of the first changes he urged was advancing the date that the bulls went out with the cows from 30 June to the second week in June. Because weight for age was becoming an extremely important factor in marketing, Chance approved this move. The cows dropped their calves earlier the next spring, so that by the time the flush of green shoots increased the cows' milk production, the calves were big enough to benefit. Those three weeks made such a difference that by November weaning they were much bigger and fatter than ever before at that time. It was the practice to wean at Minnie Lake and drive the calves to the Home Ranch in one large bunch. Whereas cows and calves are difficult to drive, calves alone, being so skittish, are even harder.

The calf drive was ready to leave on a cool November morning in 1953. The eight-month-old calves started to filter through the open gate, carefully guided by Mike's cowboys positioned at point, on each flank and at the drag. The healthy calves took some time to be moved in a uniform direction, but gradually the point established itself. Then one of the cowboys made the fatal mistake of running ahead to open the next gate. In an instant there was bedlam. The calves at point, startled by the sudden action, veered away at a run. The rest of the herd followed behind at a stampede pace. Try as they would, the cowboys could not stop them.

The stampeding calves took out 300 yards of the first fence they met, 200 of the second and 100 of the third; they were getting strung out. The tragedy was that these calves had such good fat cover that they heated up very quickly. Many lay down, exhausted by the exertion and their own body temperatures. Fifteen never got up again.

The losses appalled Mike. From then on, he and his cowboys drove the cows and calves down to the Home Ranch together, initially in small groups and later in two big drives of 2,000 to 3,000 each. Then the cowboys weaned. Despite the Spencer and Ross decree, Chance had the Minnie Lake fences rebuilt.

Frank Ross's wartime effort, his many philanthropies, plus his directorships and board chairmanships in more than 20 public companies won him the appointment of lieutenant-governor of British Columbia for the period 1955-60. Lieutenant-Governor Ross entertained more members of the royal family and more well-known dignitaries at Government House than British Columbians had seen during any of the prior five-year offices. Honours breed further honours and awards, and Ross's lifework and benefactions attracted many as the years went by. Six cities and towns appointed him freeman; universities conferred on him four honorary degrees; the Boy Scouts awarded him the Silver Wolf; he became a Knight in St. John of Jerusalem, Newsman's Club Man of the Year for 1959, Honorary Colonel of the Canadian Scottish Regiment; and Victoria named him their Good Citizen of 1960.

When fire razed Government House one chilly April morning in 1957, Frank and Phyllis Ross were undaunted. They rented a suite in the Empress Hotel at their own expense for the two years that the new Government House was under construction. The Rosses left the antiques with which they furnished this official home to the province in perpetuity. Frank Ross was proud to serve the country which had treated him so well.

It was through Ross's position as lieutenant-governor that one of Brian Chance's headaches was removed. Each year Chance attended the Kamloops and Calgary bull sales to make the annual purchases necessary to keep Douglas Lake's 200-bull herd young and vigorous; on an average he bought 25 Herefords and 5 Shorthorns each year. These were bought a few here and a few there; none of the Alberta or B.C. purebred breeders produced a sufficient number to allow Chance to buy them all from one man. Colonel Spencer raised bulls at his Earlscourt Farm, but not in adequate numbers.

In 1953, Spencer and Ross had started pushing the ranch towards a goal of a 4,000-head calf crop; this naturally meant that the cow herd must be increased and more bulls found. Because decades of meticulous selection had greatly improved the Hereford breed genetically, the need to crossbreed was disappearing. As well, the demand for grain-fattened beef placed Shorthorn cattle at a disadvantage because they had the ability to fatten—even overfinish—on grass before they ever reached the grain feedlots. The Colonel preferred Herefords, and the Shorthorn cattle were removed from the herd, requiring these bulls to be replaced also.

Chance began looking in Canada and the United States for a purebred breeder who could supply annually up to 60 young Hereford bulls with sound feet and legs, good bones, and the constitution to get around in rough country and still produce a high percentage of growthy calves. Such substantial purchases would allow heavy culling from those bulls when they reached four years old. It was a difficult search, but he felt that he had found the right breeder in Curtice W. Martin of Montana. The Douglas Lake owners, however, were not so enthusiastic, especially Spencer, who would have preferred to see the business go to a British Columbian. But none was available.

A guest of Lieutenant-Governor Ross, American Ambassador Stuart had a background in cattle feed and feedlots. In discussing his cattle ranch, Ross mentioned his manager's suggestion that they buy bulls from Curtice Martin and Stuart enthusiastically endorsed the idea, for the fame of the Beau Donald bloodline of Martin's ranch had spread throughout American ranching circles. Ross quickly advised Chance to settle arrangements with Curtice Martin.

Unlike the packers who had bought grass-fed beef for the flesh under the hide and never complained about red necks (animals with no white hair around their necks), or line backs (animals with a white line from poll to tail), or even a mottled face, the commission men buying for the feedlot operators insisted on uniformity. As feedlots were buying all of Douglas Lake's annual production from 1956 on, Chance looked to the Curtice Beau Donald line to provide regular colour, height and conformation. Over the years, the offspring met Chance's expectations and buyers reported that Douglas Lake animals carried more weight for age, and converted hay and grain more effectively.

From 1957 on, Spencer and Ross became disenchanted with the cattle business. "You have to figure that Douglas Lake needs an income of approximately $500,000.00 per year," commented Ross, referring to the amount required to meet operating expenses. He and Spencer were afraid they were "going to suffer very substantially because we will not have the volume to sell to cover our expenses" for the year ending 30 April 1958.

As Chance tried to point out, Douglas Lake had actually served Spencer and Ross well.

For purposes of making a capital recovery, 7902 head of cattle were disposed of for the most part between September, 1950 and November, 1951. This figure covers the saleable cattle which normally would have been shipped out in the course of the Company's financial year ending April 30th, 1953 . . . .

. . . while making substantial provision for the extension of the Harry's Crossing project, and paying operating costs etc. for the years ending April 30th, 1951 and April 30th, 1952, the sum of $700,000.00 had been applied against the Bank loan then being carried by the ranch.

This put the Company short of one year's income.

Since that time I have had to strike a balance between building up the herd again to the point now required, and yet sell enough cattle to pay expenses and show a profit.

A declining cattle market since the spring of 1952 in an otherwise expanding economy has contributed towards making this a slow and difficult undertaking . . . .

Despite Ross and Spencer's misgivings, the year ending April 1958 showed a $127,823 profit. An additional $500,000 had accrued to Douglas Lake since 1952 from selling all the merchantable timber—Douglas-fir, jack pine, and ponderosa pine—to various logging companies in the valley.

At the end of 1953, Ross had requested an annual calf crop

208

of 4,500. He reduced this request to 4,000 the following June after consulting with his auditors, but it was still an amazing expectation. The average annual calf crop of 2,650 in the early '30s had risen to 3,050 in the late '40s. Although it seemed impossible, this steady rise had actually accelerated in the first few years of new ownership (even while the top of the cow herd was being sold off) to an average of 3,500 calves annually. This was achieved by breeding the yearling heifers. Ross and Spencer still needed an additional 500 to 1,000 calves to cover their outlay in purchasing Douglas Lake Cattle Company. Incredibly, the ranch met the demand and the average annual calf crop in the years 1955-58 inclusive reached 4,200.

In February 1958, because of the rising cost of living, the poor cattle prices, and the fact that yearlings instead of two-year-olds were in demand, Ross enlarged his request to 4,750 calves. Chance decided to breed 600 yearling heifers, yet at the same time he was afraid that the additional grazing demands from the numbers of cattle involved in a retained "brand" (branded crop) of 4,750 calves would be more than the ranch grass could support. That summer of 1958 proved it was possible, and Chance agreed to maintain the breeding herd at 5,500 cows and two-year-old heifers.

In the eight years that Spencer and Ross had owned the ranch, profits had been poor: only in their initial years were the profits in line with their other investments. They had gradually learned what other ranchers already knew well, that the ranching business, which is so dependent on variable factors, brings poor returns for dollars invested. The joys of owning and improving such a superb spread as Douglas Lake; of being involved in a physically demanding and basic enterprise; of riding through the wide range as dawn breaks, passed them by. Only to those who worked on the ranch were these sufficient rewards. Disillusioned, Spencer and Ross sought purchasers for the ranch, which had almost doubled in value since 1950. Their ages—Spencer's 77 and Ross's 68—were added incentive to sell.

The two financiers asked Charles N. Woodward, another Vancouver businessman, if he would be interested in buying

Douglas Lake. He had been up twice hunting—once when he thought he was poaching and once at the invitation of Spencer and Ross—but on neither occasion had he looked at the operation. He came out again and looked around more seriously, talked to Brian Chance, and returned to the coast to discuss the financing with a stockbroker friend, John J. West.

Barely six months after the first overtures, Douglas Lake transferred to two new owners, Woodward and West, who incorporated Douglas Lake Cattle Company (1959) Ltd. on 24 June 1959. The four men signed the final documents of transfer on 20 July 1959; the purchase price, which included a mortgage, was $2.6 million.

During their nine-year ownership, Spencer and Ross had made some incredible demands upon Douglas Lake's land and cattle, manager and staff. In most instances, the response was positive, indicating the stability of the operation. However, apart from adding 17,800 acres to the deeded land, and building up the potential production of the cattle herd, their business strategies slowed progress elsewhere and even hurt the ranch. This was especially true of their negotiation of the half-million-dollar timber sale which lingered on for more than 20 years because the contract had not stipulated the actual volume sold or the logging methods and cleanup practices required.

The sale of the ranch marked the end of Spencer and Ross's long-standing business relationship, for the Colonel died the next year. The lieutenant-governor died 12 years later.

CHAPTER EIGHTEEN

In looking back through the early years after the purchase and sub-sequent events I am convinced that Chunky did not purchase the ranch as an investment. I believe that above all of his other activities his first love is ranching, particularly after his long family association with Alkali Lake Ranch. His main desire was to own his own ranch, and of course the plum of them all was Douglas Lake. It came up for sale at a very opportune time. Jack West possibly had different motives, as he was primarily an investor, although also an avid shooter and excellent fly fisherman. On any basis I am sure he realized that he could not lose on the purchase.

John M. Tennant of Lawson, Lundell, Lawson & McIntosh,
barristers and solicitors to Douglas Lake Cattle Company Ltd.,
to the author 15 February 1979

Charles Namby Wynn Woodward, or Chunky, as his friends called him, in 1945 joined the department store which his grandfather had started in Vancouver 53 years earlier. When he became president in 1956 at the age of 32, Woodward Stores Limited had seven stores, and an eighth was planned. His father, Colonel William Culham Woodward, who had been president before him, died the next year.

Chunky Woodward's maternal grandfather, Charles E. Wynn Johnson, a close friend of Frank Ward, had purchased Alkali Lake Ranch in the Cariboo in 1908. Over the next 30 years he built it up to a 3,000-head operation, and Chunky had been visiting the ranch since he was three. Memories of happy childhood holidays spent riding there moved Chunky Woodward to try to purchase his grandfather's old ranch from Mario von Riedemann, the son of an immigrant Swiss who had bought it near the end of the depression. Woodward found that the charming spread set on the benchlands above the great Fraser River was not for sale.

Woodward knew little about the business of ranching, of how to make a large ranch pay. His stockbroker partner, J. J. West, knew no more about ranch economics.

John Joseph West, a fifth-generation Canadian on his father's side, had also grown up in Vancouver. A commerce graduate from the University of British Columbia, he had joined the national investment house of Wood Gundy Securities at about the time that Chunky had joined Woodward's. They met on a Cariboo hunting trip soon after. In 1948, Jack married June C. Jewell, the same year that Chunky married Rosemary Jukes. In the dozen or so years since, Jack West had risen to the position of vice-president and operations manager of Wood Gundy in Vancouver.

It was not because they wanted to be ranchers that Woodward and West bought Douglas Lake: it was rather that they were both fanatical outdoorsmen; West even assembled his own guns, carved his stocks, hand-packed his shells, and tied his flies. They saw the 75-year-old ranch as a good financial investment that could also provide some of the best hunting, fishing and riding opportunities in British Columbia. So initially they visited the ranch more as sportsmen than as businessmen. Rosemary Woodward was an accomplished rider in the English saddle, but as she abhorred dust, she suffered through the Interior weekends.

Throughout the '60s the calf crop hovered around 4,500 although the total weight of beef sold climbed upwards. Wages also gradually rose and Chance had to watch carefully the numbers of men he employed.

The ranch was feeling the effects of all Spencer and Ross's timber sales, for hunters, fishermen and hikers now had easy driving access along the logging roads to the many beautiful lakes east of the Merritt-Princeton highway. Their presence unsettled the cattle, which became too nervous to drink at their habitual watering holes. On the other hand, the logged area was sown with tame grass cover and refenced to carry more cattle.

On 14 June 1960, Brian Chance told the fish and game and rod and gun clubs in the area that the ranch would be closed to hunters and fishermen following the 1 July holiday. The reaction was immediate: "Douglas Lake's Closure Felt by Local Fishermen" headed an article in the *Similkameen Spotlight*, which prophesied that "this move by the Company will have an effect on business in Princeton over the years."

Jack West supported Chance's move.

We are quite within our rights to close off our own lakes.... The presence of an ever-increasing number of campers interferes with the operation of our business. There is no point of saying that a few bad apples spoil the barrel as we certainly are not going to afford the time or the money to sort out the apples, nor is there any way that anyone else can do it for us. Water and range land of Douglas Lake represent our only asset. Anything that interferes with the conversion into meat costs us money. If we have to divert one rider for one half hour to remedy the problems created by trespassers, that is one half hour too much.

I feel very strongly that unless we close this area irrevocably and take the roads right out by the roots and make the area completely impassable we will ultimately be shoved back by hordes of the public until we are operating a hotdog stand at the home ranch.

A week later his stance was the same.

It is essential that this be done even though it may deprive some individuals of certain benefits of trespassing. The fact remains they have the whole countryside to camp in. The Social Credit book boasts that there are now 155 Provincial parks and that more are being acquired every day.... They go on to say that through their magazine, beautiful British Columbia is receiving worthwhile acclaim from the tourist industry. It is interesting to note that in one of the issues several pages are devoted to Douglas Lake Ranch as if it were a combination English Bay-Stanley Park.

But by early the next year, public pressure had forced Brian Chance to moderate the company stance and he discussed with the Department of Conservation and Recreation the possibility of selling public access to Courtenay, Crater and Pothole lakes back to the Crown. "This is cooperation at its best between industry and the government in the best interests of the people," reported outdoors journalist Lee Straight. "The Douglas Lake Cattle Co., in its 80 year history has always been one of the pioneers in appreciating multiple-use aspects of the land. This, in an area where there are so many lakes and so much hunting that the problem will be an ever present one."

The company granted access to a number of small lakes in the Pothole country, subject to the use of designated campsites under government supervision. But this was just the first of the recreational problems that the ranch would have to solve.

Woodward and West began to take part in such ranch operations as the big fall cattle drives, and in such decisions as the one to sell all cattle that showed their Shorthorn cross bloodlines in their brockle faces, dark hides and big frames.

They also approved various land purchases: Vic Shropshire's half section along the Merritt-Princeton highway at Aspen Grove, and Norman Wade's 1,600-acre Echo Valley Ranch east of Aspen Grove. They authorized the Norfolk foreman, Fred Reimer, who had replaced Findlay Anderson, to clear more land. The willow brush that covered much of the Crowhurst Field died after a helicopter dropped the herbicide 2,4,5-T on it. Modern chemicals had made an age-old task easier. Reimer and his crew next straightened the winding Salmon River, giving it a new channel on the edge of the flat. As a result of these efforts, Crowhurst hay tonnage quadrupled.

Then early in the '60s, Woodward insisted on another change which neither West nor Chance considered beneficial to the ranch. He began a Quarter Horse breeding and training program that necessitated, right in the middle of the Home Ranch yard, a barn, a barn boss and a horse trainer. West thought the idea crazy; Chance was certain the program was more than the ranch could support. Just as Frank Ward's family had protested the polo ponies that he grazed and trained at Douglas Lake, so Chance and West opposed Woodward's Quarter Horses. But in fact, the pleasure that this operation brought Woodward was one of the few dividends he received from the company, for he plowed back all profits into improving the ranch. This in itself pacified West and Chance.

At this time, Woodward met Vacy Ash, the president of Shell Oil in Toronto and the organizer of the 1962 Ottawa Commonwealth Conference that His Royal Highness Prince Philip, Duke of Edinburgh was to attend. The Duke, who was also coming west, wanted to relax for a few days on a ranch, and Woodward agreed to Ash's suggestion that he play host to the royal visitor at Douglas Lake.

The preparations for the week-long visit were extremely

thorough. Representatives arrived from Buckingham Palace to inspect the Duke's accommodation, arrange a program of activities and learn what the countryside had to offer. Aerial photographs of the buildings and of the area were taken.

Then security men made more minute checks. The house in which the Duke was to stay was the one built years earlier for Brian and Audrey Chance. Since they moved out, it had served as the bookkeeper's residence and guest cottage. Spencer and Ross had lavished thousands of dollars on the interior, sheathing it with knotty pine and furnishing it with ranch-style maple. It was now Woodward and West's home on the ranch. A tree grew close to the window of the room where the Duke was to sleep, and the RCMP officers from Ottawa asked Chance to have it cut down as a precaution. Instead, he promised to guard it so that no one would find an opportunity to climb it.

The police searched the backgrounds and records of all 90-odd ranch employees and ordered Chance to send two men on holiday during the royal visit because they held subscriptions to an underground newspaper. The plainclothes police who were assigned to the job set up their own barracks. Chance hired an additional pair of hands for the Home Ranch cookhouse to provide meals around the clock. A two-way radio system permitted 24-hour communication.

Next, everyone concerned learned the Duke's likes and dislikes: he likes gin; he does not like whiskey, playing cards, having his photograph taken without his permission, or signing too many autographs. He does not like people feeling that they must mow their lawns and paint their gates for his visits—though the desire of his hosts to tidy up their property is inevitable. In fact, the visit became a good excuse to remove from the Home Ranch yards a very large mountain of manure and to repaint all the barns and buildings.

The Woodwards were helpful in advising Chance on the arrangements for the royal visit. Points of etiquette came easily to Chunky's mother, for her husband had been lieutenant-governor of British Columbia from 1941 to 1945 and, as the Queen's representative in the province, Colonel W. C. Woodward and his wife had had many opportunities to meet and entertain dignitaries.

During a rehearsal, Yee, the Chinese cook in Woodward and West's house at Douglas Lake, set a jar of toothpicks in front of the Duke of Edinburgh's table setting. Yee was diplomatically if hurriedly exchanged for Wong, the Woodwards' Vancouver cook, who had, as he announced proudly, "cooked for many high up diplomat before."

The royal visit went extremely well from the moment the RCMP float plane touched down on the waters of Douglas Lake on 27 May 1962, and the Duke, his bodyguard Inspector Frank Kelly, and the rest of his entourage stepped out onto a jut of scrubby shore known afterwards as Prince Philip's Point.

The Duke fished, rode, bird-watched and attended many of the day-to-day ranch activities. One day at the Raspberry Corrals he watched as Jimmy McDonnell broke a green two-year-old colt. Before climbing on, Jimmy choked the horse down, halter-broke it, sacked it out, saddled it and made it bridle-wise. The Duke was enjoying the lack of formality and turned to Brian Chance, saying, "Thank God there are no policemen around." The manager glanced around the corral at the half-dozen Mounties dressed up as cowboys and at the two powerful limousines disguised as jalopies which were kept ready for the instant Prince Philip decided to go bird-watching or fishing at Salmon Lake. A uniformed roadblock manned both routes in and out of the ranch, preventing every travelling salesman and tourist from entering the area. The Salmon Lake fish camp suffered a poor week's business. But the Duke enjoyed total freedom, or so he thought, so that a visit to Chapperon was sufficiently spontaneous that Mrs. Lawrence Graham had no time to remove her apron before shaking hands with the Duke over her garden fence. That was the way he preferred it.

The Mounties' country attire indeed disguised them well, as Jack West found out. After watching one of the ranch hands for roughly half an hour as he idled at his job, West complained bitterly to Chance. "No wonder our profits are so poor if that man is any example of the kind of labour you employ." "He," replied Brian Chance quietly, "is an RCMP officer."

Another day the Duke watched a cutting demonstration.

The Quarter Horse breed had received its name in the southern states, where farmers developed the animals for light draft work during the week and for quarter-mile sprints at Sunday races. But their ability to "cut" (separate) cattle intrigued Chunky far more, for on a ranch cowboys must regularly separate calves off for weaning, sort cattle for sale, isolate sick animals, and do similar riding. Rodeos now feature cutting demonstrations and competitions.

Three of Chunky's cutting horse friends helped in the small royal display, and the Duke joined in. The hour and a half stretched to four hours as the Duke rode each mount and tested its ability to outguess the gregarious cattle. The entertainment had a delightful repercussion. Later that fall, the Duke invited Chunky and the Canadian Cutting Horse Association to give a display at the Royal Windsor Horse Show. Eight riders and their mounts flew to Britain in 1964 for a three-month tour that culminated at Windsor. Here Chunky presented Prince Philip with a Quarter Horse filly out of one of his Douglas Lake mares. This filly, at that time only halter-broken, became a polo horse for both Prince Philip and Prince Charles.

The phenomenal success of the tour was indicated by the immense press coverage and the warmly welcoming British audiences. When a second invitation to the Royal Windsor came in 1969, Chunky Woodward and his cutting horse friends declined the offer, for they felt that another tour could only be an anticlimax after the tremendous reception of 1964.

Not long after the Duke left Douglas Lake that June 1962, one of the stableboys, Dave Batty, rode the grey mare which the Duke had used several times during his stay. The mare bucked and threw her rider clear of the saddle. Dave looked up, stunned, wondering why the Duke had not received the same treatment. "Guess you haven't got a Royal arse," quipped Jimmy McDonnell, the trainer.

This was not the first royal visit, for in 1959 the Queen and Prince Philip had spent a few days at the Dole fishing lodge on Pennask Lake. Lieutenant-Governor Ross learned something of protocol on that occasion when the entertainment he had laid on for the Queen had to be cancelled. It took her away from a washroom for longer than the maximum three hours.

In 1964, Señor Eduardo Campos came to British Columbia on behalf of CORFO, a Chilean corporation that proposed to improve cattle standards in Chile by importing 15,000 heifers and 600 bulls from around the world. British Columbia ranches agreed to supply Chile with 893 of these cattle, and Douglas Lake Cattle Company would supply 527 of that number.

Neil Woolliams, the commerce graduate whom Brian Chance had hired in 1961 with the idea of later making him assistant manager, jumped at the chance to accompany this cargo, as did Vern Ellison, the chairman and manager of the Kamloops Bull Sale. Jointly they were to be in charge of the care, feeding and health of the animals during the 27-day sea voyage on the *Clara Clausen.* It was an eventful trip for the British Columbians feeding and mucking out the cattle in such close confines and treating diseases that included shipping fever, pinkeye, bloat, footrot and finally constipation. But at last they arrived at Valparaiso with the 894 head—a cow had delivered a calf en route.

That cargo to Chile was memorable for another reason. It was the last group of cattle to travel by rail from Nicola to the coast. Ranchers in the Nicola Valley had begun trucking their cattle to feedlot owners on the prairies, and the Nicola line, so important for so many years, had finally fallen into disuse. Train service to Nicola stopped completely, the yards came down, and many ties went for fence posts.

Woodward and West introduced new policies, and stack wagons, larger tractors and better tillage equipment streamlined the farming operation. One man on a self-propelled New Holland bale wagon could put up as many as 3,000 bales—120 tons—of hay each day, thereby supplanting a stacking crew of six to eight men. Raymond Graham's sons, Lawrence, Vernon and Neil, helped pioneer these machines, putting up more hay with them than probably anyone else in North America.

This mechanization gradually increased the tonnage of hay

produced and decreased the manpower needed. It also replaced the Clydesdale teams, some of which went to Peace River ranchers; R.J. Bennett, son of Premier W.A.C. Bennett, purchased one team to use on the Eldorado Ranch in the Okanagan which he owned in conjunction with his brother, Bill, later British Columbia premier himself; and Potter's Distillery bought teams for its show wagon. The Home Ranch forge closed its doors forever.

Running the ranch with more machinery and equipment certainly required fewer men, but they had to be mechanically minded and trained. No longer did the ranch need a large pool of Indian labour to draw upon, and by the '60s just two Indian farmers—August McCauley and Joseph Charters—were on the crew, operating the big tractors at Chapperon. The Chinese period was also coming to an end: the irrigators were moving to restaurants and market gardens.

Chunky's Quarter Horse operation, now located on the far side of Sanctuary Lake, was increasingly successful. Peppy San, a stallion costing $15,000 as a three-year-old in 1962, won $20,013 and became World Champion Cutting Horse in 1967. Matlock Rose trained and rode Peppy San into the world championship title, just as he had ridden Stardust Desire into the 1966 world title. Dave Batty, who worked his way up to the horse trainer's position, and Chunky Woodward both took many trophies with the horses in Canada and in the States.

Peppy San proved to be a valuable profit maker, for not only did he breed visiting Quarter Horse mares for a substantial fee but he also bred many of the cow camp mares. This new blood introduced more "cow sense" into the working horses. Peppy San's many registered offspring continually made news in Canadian and American Quarter Horse winning circles.

In 1967, Brian Chance decided to hand over the reins. In fact, with his second wife, Jean, under doctors' care at the coast, he had been spending more and more time away since 1965, leaving the management of the ranch to Neil Woolliams. In August 1967, Chance officially retired and left for West Vancouver, where he bought a house overlooking the fifth hole of the Capilano Golf Course.

Chance had been at Douglas Lake all his working life—a total of 42 years—and for 27 of these he had managed Douglas Lake Cattle Company. He had seen three sets of owners, a world war, a changing marketplace that demanded younger and younger animals, the changeover from a labour-oriented ranch to a mechanized ranch, and the enlargement of the deeded acreage to 163,000 acres plus grazing rights, and through all this period he had given strong leadership.

The man Chance considered capable of managing Douglas Lake during the years ahead was Neil Woolliams, a new breed of manager with a university degree and modern ideas. The two owners agreed to try him out. Just as Chance and Ward had had to do before, Woolliams would have to prove himself capable of running a ranch that now covered half a million acres.

Peppy San (photograph by Orren Mixer)

Brian K. de P. Chance (courtesy of
Brian K. de P. Chance)

Lieutenant-Governor Frank MacKenzie
Ross (photograph by Yousuf Karsh,
courtesy of Mrs. F. M. Ross)

Colonel Victor Spencer (courtesy of
Victor V. Spencer)

E. Neil Woolliams (photograph by Artlite Studio)

C. N. W. Woodward

H.R.H. Prince Philip and Stan Murphy in West Wasley (courtesy of Stan Murphy)

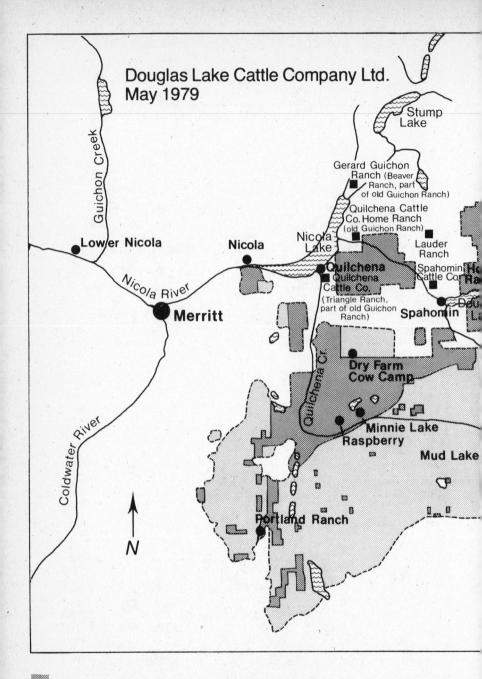

Douglas Lake Cattle Company Ltd.
May 1979

Stump Lake

Gerard Guichon Ranch (Beaver Ranch, part of old Guichon Ranch)

Quilchena Cattle Co. Home Ranch (old Guichon Ranch)

Nicola Lake

Lauder Ranch

Spahomin Cattle Co

Lower Nicola

Nicola

Quilchena
Quilchena Cattle Co.
(Triangle Ranch, part of old Guichon Ranch)

Spahomin

Merritt

Nicola River

Guichon Creek

Quilchena Cr.

Dry Farm Cow Camp

Minnie Lake
Raspberry

Mud Lake

Coldwater River

Portland Ranch

N

deeded land   ------ ranch boundary

grazing permit   ■ Headquarters of neighbouring ranches

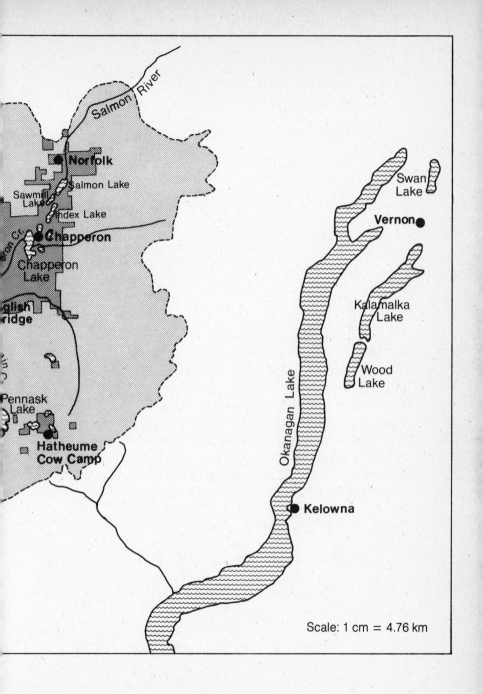

Salmon River

Norfolk

Salmon Lake

Sawmill
Lake

Index Lake

Chapperon

Chapperon
Lake

glish
ridge

Pennask
Lake

Hatheume
Cow Camp

Swan
Lake

Vernon

Kalamalka
Lake

Wood
Lake

Okanagan Lake

Kelowna

Scale: 1 cm = 4.76 km

CHAPTER NINETEEN

It is my belief that the only way a large ranch like this one can fight the trend of rising taxes and other fixed costs is to intensify and produce more beef, and to diversify to take advantage of other natural resources that can become more income producing. . . .

Further increases in cattle numbers here involve a continuation of a policy to do more fencing and more farming than are required to simply maintain present productivity. All other things being equal, there is no reason why this ranch cannot be running about 12,000 through the winter by the year 1977. . . . It must be done if the ranch is to be profitable.

E. Neil Woolliams to Board of Directors,
22 August 1968

When Neil Woolliams first hired on at a ranch, in the summer of 1953, it was not entirely with his parents' approval. Through an uncle, he had met the manager of the Guichon Ranch in the Nicola Valley, Gerard Guichon, and went to Rockford to help with the haying. The hours were long and the work arduous—driving team for mowing and raking, and stacking bales by hand—but the tall 15-year-old revelled in it and could barely wait to try his hand at cowboying in the fall. However, his uncle had asked Gerard to "work the hell out of him; he must go back to school." So when it was time to move out to the cow camp, he was told that there was no more work for him and he returned to finish his schooling at Summerland in the Okanagan Valley.

Not until he had only one year left of commerce and business administration at the University of British Columbia did Neil come back to the Nicola Valley. This time he helped put up Douglas Lake's 1960 hay crop.

Over lunch at the Georgia Hotel in Vancouver in 1961, Brian Chance asked Neil if he were willing to come to Douglas Lake for a three-year trial with the possibility of afterwards becoming assistant manager. Neil had received other offers with better pay, but ranch life had intrigued him and he accepted. His first job on this occasion was helping in the office while the accountant holidayed in Europe.

Towards fall, Neil moved out to Aspen Grove to ride for Joe Sledge. He soon wrote to his parents:

Joe is responsible for about 3,500 head of cows each summer. Mike Ferguson, the cowboss, brings them over from Douglas Lake, Chapperon, etc., in the spring and Joe and his crew pick them up at Courtenay Lake. The majority of the cows have calved by this time, as it is now well into June or maybe even into July.

You can understand how it would be hard to turn the cattle out in the summer. These cows must leave fine spring and fall ranges (open) and plod onwards to timbered ranges of higher elevation. Consequently the most men are needed in June & July for the turn out.

From this time on, the crew that is retained, (usually Joe and one or two more men) pack salt, fix fences, and keep moving cattle back up into the hills that drift down lower again. On this ranch, we are still living partially in the time of open range. For example, the 'field' across the road (east side) locally called the 'One Mile,' has only a fence on the highway side. One could go straight across to the Okanagan without hitting a fence. Natural barriers are all that stop cattle from drifting over to the O.K. In fact, with the logging roads that are now being put in, some cattle have been found over at Peachland in the fall.

You can visualize the problems of trying to locate cattle... should they decide not to come down with the ensuing cold weather. And I guess there are always a few that just won't move.... It went down to 20° [Fahrenheit] last night and this a.m. it snowed until about 10:00. This sort of weather should bring a few of them down. From the Portland here we push the cattle up towards Merritt; they get into the open and then start drifting towards Minnie Lake.

I did think I might make it home this weekend but when you are working seven days a week, there isn't too much time for travelling. However, so far I'm not missing out on relaxation. Last night Dick Boege, Victor Goethe and I went to town [Merritt] and bowled a few games in the alley that hasn't been open a week. I have gotten into a mixed league with Dave and Elisabeth Doumont....

In the summer time a cowboy must drive cattle when it is cool, so he must get up at 2:00 or 3:00. Now, since it is 'cool' all the time, and since it doesn't get light until 5:30, there is no need to get up before this time. We are usually away from here by 6:30. One of the three of us must 'wrangle' or 'jingle' the horses, so the wrangler is up slightly before daylight. I guess the horse pasture is 40-50 acres at least and it sometimes takes a while to locate the horses. One horse is always kept in for wrangling.

For the next couple of years, Neil rode with George Sicamen, a Spahomin Indian with 45 years' experience in the saddle; Sicamen had been just 12 when Joe Coutlee added him to the Douglas Lake branding crew. Neil learned much about the country, cattle and horses from this great Indian cowboy and his half brother, Andrew Jackson.

The brothers' ribald jokes and fun-loving tales fostered in Neil a great fondness for the Indians, and he began to comprehend the interwoven lineage of reserve families. He even gleaned a smattering of the Okanagan language, enough to enable him later to deal with the Indians in their own glottal tongue. Chance and Ward had only mastered the more anglicized Chinook. This facility helped Neil's relations with his native neighbours.

Chance moved Neil all over the ranch to give him a thorough background: foreman at Mud Lake, rider at Norfolk, foreman at Minnie Lake, storekeeper, accountant, cowboy on the Westwold drives, and even cook while his leg was in a cast. He became assistant manager when he returned from taking the cattle shipment to Chile.

Cattle went to Westwold for the last time during the winter of 1966-67 thus ending a 70-year pattern. Hay purchases that winter totalled $103,871, nothing out of the ordinary, but the next year, with no Westwold hay to buy, hay purchases totalled only $62,000.

The permanent camp at Harry's Crossing closed down in 1967, Raymond Graham sending the hay crews out from Chapperon. The use of pickup trucks and stack wagons made this possible. One by one the better buildings were moved to new locations and uses.

That same year the number of calves branded reached 5,000 for the first time in Douglas Lake's history. The number of pounds of beef sold is a better measure of production, however, and that had risen to 3.4 million, compared with 2.7 million in 1915, 3.1 million in 1934, and 2.6 million in 1963.

It was in July 1967 that Brian Chance retired, recommending Neil to Woodward and West as their new manager. They agreed, and Douglas Lake's fourth manager was the youngest: he was just 29 years old.

A number of Neil's ranch foremen arrived around the same time as he stepped into management. When Joe Sledge retired at the end of 1967, Gerry Matheson, an Alberta cattleman who had managed several large ranches, including the 6,000-head Waldron Ranch, became foreman at Aspen Grove. Orval Roulston had become foreman for the Hatheume area, using the old Chapman buildings and opening up the country until it attained a 1,600-head summer grazing capacity. Morris Roth, foreman of the Home Ranch haying operation, stepped down to Arno Nielsen, a young rancher from Brooks, Alberta. For some years, Arno attempted to run Minnie Lake from the Home Ranch, in the way Raymond ran Harry's Crossing, but the acreage under irrigation and the distance from the Home Ranch made this undertaking impossible. Ken Allen, a hard-working farmer from North Wales, succeeded Fred Reimer at Norfolk.

The key men of longer standing included Rolf Timmermanns, the German bookkeeper; Raymond Graham, the tried and true farmer at Chapperon; Stan Murphy, a New Zealand stockman with experience in sheep, in charge at the Dry Farm; and Mike Ferguson, cowboss.

Mike was fast on his way to becoming a legend in his own time, as Coutlee had been. In addition to being a leader whom the cowboys could look up to, a keen judge of an animal's health, weight and finish, and an accurate assessor of the quantity of feed in a field, Mike could recall the life history of particular cattle. From daylight to dark, he could sort thousands of cattle within a set of yards until each one in a pen was identical to the next in conformation, finish and weight: just what the packers ordered. His eagle eye could spot a four-point buck on a sidehill of fawn-coloured rocks at 400 yards as fast as he could spot a stray among a herd of "hundred and elevens" returning from summer range. Ranch audits and more demanding banking regulations required accurate cattle counts each fall, an endless job suiting only the calm and composed. One by one into the ten thousands, Mike unerringly counted the animals to give a precise tally.

A number of the older ranch hands at first viewed the youthfulness of their manager with some misgivings but Neil Woolliams soon earned the respect and trust of all his men.

He established two basic goals for the years ahead. The first was to produce and market more beef each year, which meant increasing the herd size from the 10,000 figure that Ward had accepted as a workable size. Before that could happen, however, new fences would have to divide the land to better advantage. Also, Douglas Lake would have to increase its hay production from 7,000 to 8,000 tons per annum to 11,000 to 12,000 tons if it were to be self-sufficient. The second goal was to capitalize on the other natural resources of the property by selling more timber, harvesting Christmas trees, and developing and leasing fishing camp sites.

With these two goals set, came a determination to do the work with a smaller labour force of responsible men and women. This trend had already started, for the 90 full-time men employed in 1963 had become 64 during 1968. Beef prices to the producer had not increased much in 15 years, and it was necessary to do the same work with fewer men and more machinery.

The purchase of horse trailers for the cowboys and more pickup trucks for both the cowboys and the farmers improved their daily productivity. The later expansion of radiophone communication would also save many hours of travelling.

The Home Ranch country store was converted to self-service and its staff of three whittled down to an accountant and a storekeeper. Soon the only remaining signs of the once large crews were the number of superfluous horse barns and outbuildings and a Home Ranch rhubarb patch as large as two branding corrals. Meanwhile, the two-man machine shop increased activity until the place hissed with the steam cleaner, flashed with the welder and clanged with metal on metal, day in and day out.

A year after Neil took over, Jack West died. Chunky Woodward bought the outstanding 20 per cent interest in Douglas Lake from Jack West's estate, and became sole owner. However, the increasing capital purchases and operating expenses tended to make him a little more cautious than before, especially since income was starting to decrease. So before increasing the fencing budget, he wanted facts and figures to show that the ranch's cow-calf operation would benefit. Neil was asked to find similar proof that a new sprinkler system would increase the hay and green feed output. The studies were done and, year by year, more improvements went ahead.

When the last of the Chinese employees, Mao Fung, the Home Ranch gardener, and George Leong, the Home Ranch cook, left in 1968, Arno Nielsen moved some of the shacks that Chinamen had used over the years and was amazed at the quantity of miniature opium bottles hidden underneath. In their 80 years at Douglas Lake, the Chinese had remained aloof and apart.

Their departure left a void that Neil spent over a year filling. Six cooks and six flunkies came and went in the space of 15 months, one couple staying only long enough to cook and serve one and a half meals. Believing that a contented cook makes for a contented crew, Neil began to despair. Then the rotund and efficient Vera Suttie strode into the cook- house in August 1969. In between peeling mountains of pota- toes and tossing around 20-pound roasts, she repainted the cookhouse inside and instituted Saturday night parties at which she was her own bouncer. At one staff Christmas party, the Amazon proved her strength by severely testing two of Chunky Woodward's friends—dog trainer and hunting cohort Ken Gibson, and Canadian All Star football player Tommy Hinton—at arm wrestling.

The year 1970 was busy. An ambitious three-stage plan for improving the irrigation supply at both Chapperon and the Home Ranch began with the repair and expansion of Sucker Lake Dam, a reservoir in the hills behind Chapperon. The second stage in the plan was the drainage of Big Meadow, the ranch's first natural hay meadow, which had at one time stored water as a reservoir. By draining this area, the farm crew also increased Chapperon's hay tonnage. The third stage, not carried out until 1975, was the damming of Chapperon Lake. This came at a time when the Indians at

Spahomin had incorporated Spahomin Cattle Company. The new company plowed up and cut hay from Ginny's Flat, one of the Indian reserves at Chapperon that had always been range. Their water came from the increased storage capacity brought about by damming Chapperon Lake. The remainder of the increased capacity flowed down to the Home Ranch for a steadier source of water for first and second irrigations on all land north of the Upper Nicola River.

Another big irrigation project began around 1970. In addition to regular summer work, Arno Nielsen's farm crew levelled the 350-acre Sabin Flat and laid pipes underground. Concurrently, they dammed a small slough and waterway in the hills high above, creating a lake capable of storing 1,135 acre-feet of water. When the land was eventually ready for seeding down some years later, and the gravity water supply from the newly created Sabin Lake was connected with the wheel move irrigation system, it became one of the most efficient pressure systems in the province.

Also in 1970, the B.C. Beef Cattle Growers elected Neil to the position of president of their association, so that he maintained the tradition whereby each manager at Douglas Lake became involved in the ranching association.

Neil's marriage to the author also came in 1970. I was a physiotherapist from Cardiff, Wales, and had met Neil through my sister and brother-in-law, Elisabeth and David Doumont, who were on a mixed bowling team with him in 1961.

The previous year Chunky Woodward had moved into the $200,000 open-plan, lodge-style ranch house built for him west of the 3,300-foot airstrip at the Home Ranch. This was a much admired bachelor's pad, Rosemary and Chunky having recently divorced. Locally weathered barn board from the barn built by Chief Johnny Chillihitzia gave the house an immediate atmosphere, while a blond cowboy ghost that various guests believed had accompanied the barn board later gave many candlelit dinners a fascinating topic of conversation. Picture windows on either side of a stone fireplace threw light on many wildlife trophies, and revealed a view of bunchgrass stretching from the back door down a steep incline to the Home Ranch buildings clustered around Sanctuary Lake below and Douglas Lake's waters beyond. A housekeeper's cottage nestled out of sight towards the river.

Woodward's move left vacant the two-storey house built for Brian and Audrey Chance in the '30s and used successively by various bookkeepers, by Spencer and Ross, and by Woodward and West. Neil and I moved into the house with its million-dollar view of Sanctuary Lake and the gentle range beyond. We delighted in watching and listening to the Greater and Lesser Canada geese that reared their young along the shoreline each year.

Although city bred, I slowly accepted summer dust and mosquitoes, freezing winters, and Neil's erratic and long hours. Though at first awed by the immensity of Douglas Lake's landscape, I gradually adjusted to the sweeping range-land marching with fences, the summer grass burnt into at-first-monotonous creams and sepias, the bleak look of winter, and the interminable mud of spring. Then I joined the other Douglas Lake inhabitants in their love for and pride in the great ranch.

The poor marital track record of almost every wife of a manager or owner of Douglas Lake Ranch, starting with Jennie Douglas, did not worry me long. Modern-day facilities, including land phones, radiophones, freezers, cars, trucks, 50-mile-per-hour highways beyond the Douglas Lake turnoff, and an increasing number of other women living at the Home Ranch and elsewhere nearby, removed the dangers of loneliness. The ability of crew members to take holidays in Hawaii, Australia, and Britain, and of farmers and cowboys to hire on from such places, made the ranch feel more cosmopolitan and less isolated. By typing Neil's many letters, I was able to ease his burden, and they kept me informed of the daily business.

Though Neil happily switched from eating in the cook-house to eating at home, the move was not an unqualified success. Having spent one entire day in my kitchen chopping lean beef and mounds of crisp vegetables for an exotic Chinese dinner, I could not understand the obvious lack of appreciation on the part of our guests: Joe and Molly Lauder, and Gerard and Ruth Guichon. When Neil explained later that perhaps their many years of eating nothing but Chinese food had put the men off such cuisine forever, I resolved to concentrate on ranch fare—homemade bread, roast beef and vegetables. I also resolved to learn more about my new home.

The marketing of cattle changed fundamentally in 1970. For the convenience of the buyers, Nicola Valley ranchers began to sell their cattle within one day at on-site auction sales, and there were two that year. The first annual Nicola Valley Yearling Steer Sale was organized in conjunction with the B.C. Livestock Producers Co-op. On 9 September 1970, buyers came from across Canada to bid on the 2,000 Reputation Hereford and crossbred yearling steers offered at Douglas Lake Cattle Company's Dry Farm, Gerard Guichon's home ranch, and Quilchena Cattle Company's home ranch. (Guichons had divided their ranch in the late '50s between Gerard Guichon, who acquired the old Beaver Ranch, and his cousin Guy Rose, who acquired the Quilchena portion.)

The other on-ranch auction sale was the Reputation Calf Sale. Again, co-operation between the valley's ranchers was the key to success. Douglas Lake sold its 1,200 heifer calves one day after Joseph Lauder sold 100 pregnancy-tested cows and on the same day that Urban Berglund of the Berglund Ranch southeast of Nicola sold his 200 steer calves. Douglas Lake's 1,000 steer calves sold later yet, in December. The response was tremendous and the next year a $20,000 set of yards and weigh scales at English Bridge made the yearling steer sale run so smoothly that the auctioneer, Bud Stewart, was able to sell Douglas Lake's 1,025 head in 17 minutes.

By 1972 the yearling steer sale had become the B.C. Yearling Steer Panorama Week offering 4,000 head. Ranchers at Ashcroft, Cache Creek and Williams Lake had caught on to the idea, and the buyers kept busy for three days travelling from auction yards to ranches and back to auction yards. Cattlemen termed it Canada's biggest annual sale of feeder steers, a price setter for the rest of the season.

For Douglas Lake, the annual Reputation Calf Sales were less financially successful. In order to sell the eight-month-old calves in certainty that they would arrive at their destinations in good shape, Douglas Lake had undertaken an intensive preconditioning program. Three weeks prior to the sale date, the cowboys weaned the calves and gave them a multiple shot for blackleg, malignant edema, pulpy kidney, tetanus, gangrene and other bovine diseases; they gave each a million units of Vitamin A, plus D and E; then they injected them intranasally against shipping fever and Infectious Bovine Rhinotracheitis. The calves then went on hay, so that by the sale day they had recovered from weaning and were healthy. However, by 1972, Neil stressed during an interview that although the returns from these sales had pleased him, "buyers still aren't allowing a premium to offset all the costs of preparing animals for such a sale."

The glamorous simplicity of life on Canada's largest ranch increasingly attracted publicity. Journalists, authors and television interviewed Neil and Chunky on the achievements of and plans for the ranch. Newspaper photographers captured the gaunt, hawklike features of Les Harris, the Australian fencing foreman who owed his many broken bones and his three near escapes from death to breaking wild horses in Australia and riding broncs and bulls at the rodeos in Canada. The National Geographic put Orval Roulston's wind-tensed face and an evening view of the Home Ranch between the covers of a ranching book. Weekend supplement magazine writers assembled stories of the fall cattle drives and weaning. A television crew from France spent over a week putting the cattle drives of 1972 onto film for French audiences. By midway through the shoot, the cattle were so nervous of whirring cameras and their urgent attendants that they spooked whenever they crested a hill. One Canadian Broadcasting Corporation crew who captured the spirit of Christmas at Douglas Lake achieved that faraway look in a cowboy's eye by getting him to count the stock in a pen.

One publicity director solicited the ranch's assistance in making an advertisement. A veteran rancher, Louis Stewart of Spahomin, contracted with Neil to build the corral required for the ad, in the shape and scale of a Boeing 707; the ad would illustrate that Pacific Western Airlines could fly anything anywhere. Louis built the rail fence to keep the several hundred heifers from getting out, but never expected the couple of bulls which almost spoiled shooting day when they tried to get in. Within days of the ad's running on television, PWA stopped freighting cattle and cancelled the ad. Louis took down the fence, and the ranch reused the posts and rails.

Artists such as Peter Ewart, Jack Lee McLean, Hugh Monahan, Geoffrey Rock, Ronald Woodall, John Schnurrenberger, Harold Lyon, and Gail McCance visited to photograph and paint: a snowy cabin lit from within; mallards settling on a Chapperon hayfield; Chunky astride Peppy San; Mike and his mare chasing cattle out of some brush; any cowboy following a herd over deadfalls; buckets by a barn. Within frame after frame, rugged Douglas Lake cowboys branded calves, drank coffee, rode, cleaned their castrating knives, grinned, and looked Western. The articles, films, and publicity media all stressed the romantic aspect of cowboying, as did the one book written about the ranch in the '50s, and their creators almost ignored the equally important undertaking of farming.

Douglas Lake's increasing fame attracted many followers of the 1970s back-to-the-earth movement. Barbers, hippies, ophthalmic assistants and plumbers hired on as farmers and gardeners. University dropouts and graduates enjoyed washing their socks in the river and, at the end of the day, swilling their faces and hands in the enamel bowl outside Hatheume's cabin door. A racing jockey, Keith Smith, recounted with glee the dozens of pack rats that he and his wife, Sandy, had shot around the Mud Lake camp one summer he was foreman there. While Keith rode and roped, he gradually memorized the name and pedigree of every Douglas Lake horse.

Ex-RCMP officer Barry Wallace was hired in May 1974, though he had little ranching experience, to fill the new position of range patrol. The previous summer's losses around Aspen Grove of 160 more calves than normal made this appointment essential. Ranchers accept up to 2 per cent losses, attributing them to predators, mud holes and other natural catastrophes; more than that indicates rustlers at work. This was another of the problems that came hand in hand with the opening up of ranges through logging road access.

Wallace's unmistakable presence brought success, for the following year losses dropped back to previous levels. He broke his leg while roping an animal early on and, like many another cowboy, champed at the bit until his doctor allowed him to climb into the saddle again. While he recuperated, the ranch's insurance policy that Neil had improved to provide coverage both on and off the job took care of his wages, just as it had done for several victims of tractor accidents, rodeo accidents and heart attacks.

Developments in loose haymaking and the bulk bale had sufficiently intrigued Neil in 1971 to send Arno Nielsen and Raymond Graham, the two farm foremen on whom he relied the most, to Alberta to see in operation the Hesston stack wagons that could chop, blow in and compact windrows of mown hay into three-ton hay loaves. Far fewer men could perform both the harvesting and feeding out of hay using these machines than with the old balers.

The first stack wagon arrived at Douglas Lake in 1973. Since 1967, when hay production had been 7,221 tons, the summer crops had increased to 9,533 in 1972. Every fall, Neil went to each of the four farming operations to scale the hay and work out the tonnage. Then back in his office he would make plans that detailed which herd of how many head was to winter where and for how long. In 1973 he went from the Home Ranch to Chapperon to Norfolk to Minnie Lake, scaling hay with growing trepidation: the summer drought had had a detrimental effect on the hay tonnage. It was down to 7,130 tons, lower than the 1967 level. Neil explained to Woodward that because of such a tight feed situation, a two-part solution was necessary: the ranch should ship in more grain pellets than normal from the prairies, plus pregnancy-test the entire cow herd and sell the dries before winter. The pregnancy-testing, though taking the cowboys half a week at a busy time and causing the veterinary bill to rise, was so worthwhile that it thereafter became part of the annual routine.

The costs of purchased grain and hay since the last herd

went to Westwold had been decreasing quite satisfactorily. However, during that winter of 1973-74, a total of $253,000 went to purchase hay and grain, five times more than Chunky and Neil had budgeted the previous year. Another linerload of pellets had to arrive almost every day to keep the cattle fed, and the feed companies involved could hardly keep up with the pace. It was a nightmare of a winter, of the kind that Greaves, Ward and Chance had all faced in years past. But at last the winter was over, and the next summer's crop using the new haying machines hit 11,000 tons—another first.

Whenever the ranch reaches a goal, another is set to take its place. So it was with the hay tonnage, for a herd made up of younger animals, 50 per cent of which are pregnant, needs more hay to withstand winter's rigours than does a herd from which yearlings, two-year-olds or older cattle are going to market. It also happened that the spring, summer and fall cattle carrying capacity of the ranch was increasing, for a variety of reasons. The cross-fencing program that Woodward had approved four years earlier was making better use of higher elevation grasses on spring and fall ranges by stopping the cattle from drifting down before eating the grass there. Reseeding logged-off areas back to tame grasses improved the quantity of summer grazing. Whereas in the mid-'30s Douglas Lake had annually purchased sufficient "animal unit months" from the Crown to summer 6,000 head for four months, the ranch could now find enough grass for 7,000 head for four months in the same ranges. To avoid losing the Crown grazing permits through understocking these improved ranges, Douglas Lake began to maintain a mature herd of 11,000 head which in winter demanded at least 11,000 tons of hay.

In order to increase calf crops at Douglas Lake, a program of breeding the yearling heifers started in the '70s. Only during Greaves's management and twice while Brian Chance was manager had this happened before. By 1974 the number of heifers bred had reached 750, and a Maternity Barn housing 54 individual pens and a labour room stood next to the yards at English Bridge to allow 24-hour surveillance of the heifers during calving. But disaster struck in the form of a calf scour epidemic. Once calving was over, 20 per cent of the heifers' calf crop had died of this infectious diarrhea.

Dramatic measures were necessary to prevent a repetition of the disease the next year. The Hereford heifers, 1,200 on this occasion, went with the Black Angus bulls in early June so that they would start to calve three weeks before the cow herd. Thus the cowboys already on the ranch could calve first the heifers and then the cows. (The Black Angus cross calves would be smaller and sturdier than straight Hereford calves, and thus make the heifers' first calvings easier and more successful.) In late September, a Kamloops veterinarian, Paul Christiansen, with the help of the cowboys, pregnancy-tested the 1,200 bred heifers and 6,000 bred cows; the dry heifers received one more chance, whereas the dry cows were sold later that year. In February 1975, the cowboy crew cleared the snow out of the watch pens and laid down clean, dry shavings in preparation for the calving, which began later that month.

A team of graduating veterinary students from the Western College of Veterinary Medicine at Saskatoon then arrived to learn the practical side of calving and to assist the cowboys in calving out the heifers and later the cows. A more rigid surveillance of the heifers and a faster turnout of the mothered-up pairs onto spring range were now possible. The vet team had hoped to study calf scours at close quarters that first year, but because of the improved management techniques, only half a dozen isolated cases occurred. Even so, the practical experience at Douglas Lake became an annual affair for the vets from this college.

A world buildup of cattle inventories, and high grain prices brought about a disastrous drop in 1974 cattle prices. Pre-conditioned steer calves that had sold for over 70 cents per pound in 1973 brought only 40 cents the following year. The revenues from sales at Douglas Lake and elsewhere fell far below the cost of production and that naturally affected the ranch in a number of ways.

With the sudden disappearance of a calf market, 1974 became the last year of auction calf sales at Douglas Lake, when 3,800 were sold. A cow-calf operation is more labour intensive than running fewer cows and calves and more year-lings. Also, much good grass goes into fattening yearlings and Douglas Lake has good grass ranges in abundance. The

234

rational move was to switch back to a yearling operation.

But prices were still poor and savings had to be made in other areas if the lower income were going to cover costs. Neil cut labour down to a steady crew of around 50 and put a halt to the first facelifting of the Home Ranch in 15 years. The cookhouse already sported new beige and mustard colours, the driveway entrance swayed with a growing avenue of blue spruce nursed along by aspens, but the remodelling of more of the older homes was abandoned. The budget included only the most essential capital expenses and repairs.

The merchandising mind of Chunky Woodward intervened at this time to persuade Neil to a grass-fed beef promotion. If he could produce 450 fat heifers weighing between 800 and 1,000 pounds, then Woodward Stores would do the rest. This proposal interested Buster Giles, manager of the Stump Lake Ranch, and Gerard Guichon, and they joined with Douglas Lake in filling the October 1974 order. The range-to-table beef program was so much of a success that it was almost a disaster, for staff in Woodward's meat departments worked at a frenzy to keep up with the jostling customers who—literally—grabbed at the cheaper meat, throwing packages to their friends who could not reach the counters. The meat department staff, who never expected such crowds, vowed they would organize themselves better the next time.

The B.C. Cattlemen's Association, as the B.C. Beef Cattle Growers' Association had renamed itself, reacted differently to the low cattle prices. Many ranchers who had bought in when land values were high were finding that they could not meet their bank and mortgage payments with the low returns from their calf and yearling sales. Even many of the established ranch owners were having great difficulties meeting all their expenses. The BCCA discussed approaching the provincial government for some form of income assurance, so that the costs of production could equal the revenues brought in. Some ranchers deprecated the scheme and termed it welfare.

In March 1975, Neil sent a four-page letter to every rancher in the province "regarding the full impact of our considera-tion of Income Assurance." He argued that governments of the day wanted more involvement in agricultural production and so would let consumers pay for part of their food bill over the counter and the balance in their growing taxes. He saw such a scheme as a false stimulus, whereby cattlemen would overproduce, depressing the cattle market still further. Other provinces would ask for similar schemes and the independence of the beef industry would disappear forever. Quota systems and marketing boards lurked darkly in the wings. He cited a list of nine alternatives which included the repealing of school taxes on unimproved agricultural land, an issue that ranchers had been fighting ever since this tax had been introduced.

The battle ended at the annual meeting of the BCCA that May, when a record turnout of over 300 members voted: 205 in favour of Income Assurance and 102 against. For the first time in British Columbia's history, or indeed Canada's, ranchers were asking the government for handouts. Gone were the days when the marketplace was the ranchers' only criterion. The two-thirds to one-third vote also caused a deep rift within the association along the same division.

Though in retirement at the coast, Brian Chance grieved, too. "Was it the younger men & later arrivals in the cattle business—a different generation with the present day outlook—that went for the easy way?" he asked of Neil. "It was generally thought that the industry with its past almost legendary record of independence would have put up more of a fight for the alternatives."

Income Assurance now compensated the rancher running only 300 head with the same sum that was available to Douglas Lake with its 11,000 head. And those ranchers whose principles prevented them from accepting the subsidy were playing a game of Russian roulette in the marketplace of 1975. A ranch the size of Douglas Lake had to find new ways to stay solvent and to compete in an environment of chronically poor prices and disproportionate government support.

Douglas Lake Cattle Company had almost already achieved the goals that Neil Woolliams had espoused early on. Yet the continuing low prices ate away at ranch profitability. The future required even more imaginative management.

CHAPTER TWENTY

Calling Norfolk! Calling Chapperon! Calling . . . the Home Ranch!
We've . . . got a hell of a wreck up here! I need all . . . the pull power I
can get! Do . . . you read me? I need all the pull power I can get!

Mike Ferguson on the radiophone,
1 December 1975

Communications have always been a problem for Douglas
Lake Ranch. The public telephone service connecting the
ranch with the outside world provides the Home Ranch with
just one multiparty line using crank-handle telephones that
sound a combination of short and long buzzes. Neil's office
and home and the store answer to 1F, the Quarter Horse barn
to 1K, and Woodward's home to 1W. Every call made is long
distance because it involves the Kamloops operator.

In Ward's day, tens of miles of wire and poles constituting
the private telephone system within the ranch had linked together the five hay ranches and the Morton cabin. Lines were
always breaking down just when they were most needed.
Then, Spencer and Ross introduced the first walkie-talkie
sets. For all the time that they saved, countless more hours
were lost, for the sets were forever going wrong and it was impossible to find the right specialist to fix them. Also, anyone
near the walkie-talkies could hear the conversations taking
place so that no confidential information could be passed.

By 1973 communications were breaking down constantly.
The old tube radiophones that provided the ranch with its
own communications system were costing a fair amount of
money to maintain, but even more important to the issue was
that Neil's workload as manager was getting heavier and
heavier. Often he would drive many miles throughout the day
and after supper too, to talk to ranch personnel when a radio
call would have sufficed, had the staff been equipped with

radiophones that worked well. The ranch foremen were experiencing the same difficulties.

But how do you demonstrate that improved communications save money? The installation of a brand-new solid state radiophone system with 18 sets would streamline the whole operation of the ranch by establishing good communications, but it would cost $15,000 and Neil was having difficulty justifying the expenditure. At first Woodward was against expanding the budget to include the $15,000 cost for it made little business sense. Fortunately, 1973 was a year of high cattle prices with the top price at the Panorama Sale at $73.40 per hundredweight—a Canadian record—and eventually Woodward approved the expense.

The new solid state phones were a great boon. They worked well, and the network expanded to the foremen's trucks, many homes, and those cookhouses still operating. They effectively shrank the ranch to a more manageable size. From his office, Neil could now ask Rita Garrioch, the cook at Norfolk, to lay seven extra places for lunch for the fencing crew. He could check with Mike Ferguson whether the cowboys could move some herd that day or the next. He could find out from Raymond how the haying was going at Chapperon. He could ask Orval Roulston, a two-hour drive away at Hatheume, to be at the Home Ranch the next day. In turn, the foremen could order machine parts without driving away from the job. In fact, the phone so enraptured Margie Graham (a granddaughter of the newspaper publisher, Ma Murray), who was living at Chapperon, that she kept calling her husband, Vernon, even when he was cowboying. "Margie," advised Mike over the air, "Vern doesn't have a phone on his saddle."

But there was still no way to cost the communications improvement, and the cattle prices had fallen over $30 per hundredweight by 1974.

It was a typical Monday morning, 1 December 1975, and Neil's desk overflowed with letters and files. The 1,000 head of cattle taken to Norfolk for late fall grazing were finding the snow pretty deep, and were bunching up quite badly in the south, until they collected around the north end of Salmon

Lake and put pressure on the fence across the road. Neil called Lawrence Graham at Norfolk and asked him to open the gate along the road so that the cattle could go south to the more open country. They would then stay on the bunchgrass ranges south from the Fish Lake Front, as locals still called the field beside Salmon Lake, until around Christmas before going on feed. Neil then dictated a few letters. He made a few phone calls off the ranch. Meantime the expensive radio-phone sat on his desk and clicked into activity once in a while.

"Calling Pat, calling Pat."

"Hello, Rusty."

"Can you make up a bridge four down here this Wednesday?"

"Sounds fine. See you then."

"Okay, 'bye."

Then the phone would lie idle.

"Calling the office."

"Yes, Raymond," answered Don Murray, the new ranch accountant.

"When Scotty goes in on the town run tomorrow, could he pick up some laying pellets instead of that laying mash? The chickens go back and forth over the mash and it's a powder in no time. We mightn't waste so much if it were in pellet form."

"Okay. How much do you need?"

"Let's try two tons. That should last a while."

"Righto."

The phone was silent once more.

"Calling Mike."

"Hello, Buck."

"There's a bunch of cattle walking down the lake towards White's."

"I'll be right there." There was a pause. "Lock up all them dogs, Buck, we don't want 'em spooking those cows."

Three minutes later, Buck was on the blower again. "Calling Mike, calling Mike, there's 15 or 18 head gone through the ice now."

Neil, who had heard Buck's first call, stiffened behind his desk. Those were the very cattle that should be going through the gate which Lawrence had opened. Did that mean they had gone through to Fish Lake Front and walked onto the frail ice of Salmon Lake?

Buck Greaves was a grandson of Joseph Blackbourne Greaves. After spending most of his life off the ranch, fishing commercially, Buck in his fifties moved back to live in the country so loved by his father and his grandfather. He had driven team, or fenced, or taken the odd range-seeding contract, and operated a trapline in winter. Lately, he had moved into a small cabin at the south end of Salmon Lake, near the cottage owned by Bob and Twigg White, brother-in-law and sister of C.N. Woodward.

When Mike first heard Buck's call, he was driving over Cayuse Mountain with two cowboys, Bill Brewer and Archie Charters, opening water holes and putting out salt for the cattle there. When he heard the second message, the veteran cowboss wrenched hard on the wheel, turning towards Salmon Lake. Mike raced his truck over the range; then he was on the gravel road. At last he rounded the corner that gave him his first clear view of the lake, White's cottage at the near south end, and the Salmon Lake fish camp buildings at the other end. He was horror-struck. The meagre details supplied by Buck had not prepared him for the disaster which met his eyes.

Coming from the Norfolk hay meadow at the north end of Salmon Lake, hundreds of cows had drifted down the fence line and straight onto the ice at the lake's edge. The cattle had not even noticed the open gate to Fish Lake Front. Now 300 or 400 of them were casually strolling down the ice toward White's. At the moment they were well strung out. But there were many more in the hole than the 15 that Buck had reported. The figure was closer to 50.

Mike, Bill and Archie jumped from the truck, grabbed a length of rope from the pickup box, and hurried down the snowy sidehill. As they neared the lake, they spotted another hole full of cattle quite close by. It had been one of the last places to freeze up that winter and there were still many soft spots around. The cowboys picked their way around spongy places and eventually got close enough to help the stricken animals.

The freezing water churned with 13 swimming cows. The hole they had made in the ice was too small to allow them all to keep their noses out of the water at the same time. It was like looking at so many hippopotamuses wallowing, but these cattle were in ice-cold water, swimming for their lives. A more powerful cow would get her head out a little farther, pushing her neighbours under the water as she did so. It was a constant struggle as each cow tried to stay near the centre of the hole.

The three men roped a cow and started to pull her out. But it was useless: they were not strong enough, and their feet slid on the ice. Mike ran back up the steep bank to his truck radiophone to call for more help. He could not speak at first, he was so short of wind.

"Calling Norfolk! Calling Chapperon! Calling... the Home Ranch! We've... got a hell of a wreck up here! I need all... the pull power I can get. Do... you read me? I need all the pull power I can get! At Salmon Lake. Bring ropes. And axes."

"It's that bad, is it Mike?" asked Neil from his office as he hustled into his jacket.

"More than 50 head gone through the ice. I need all the help I can get."

All over the ranch, everyone within earshot of the radiophone grabbed a coat and rushed out to start up his truck. Neil and Arno Nielsen threw all the rope they could find at the Home Ranch into the back of Arno's truck. They headed for the Quarter Horse barn to collect the crew there. Don Murray picked up everyone working in the Home Ranch machine shop. He met up with Neil and Arno, and took some of the large coils of rope. Lawrence Graham collected his crew at Norfolk and sped down the road. Neil Graham gathered everyone he could find at Chapperon and set off.

As the gravel flew under the wheels of Arno's truck, Neil was on the radiophone seeing whether anyone had missed hearing Mike's call. He called Chapperon to see if he could get a tractor out to Salmon Lake to help pull, but everyone had already left.

Even as the three cowboys waited for reinforcements, more cattle broke through the ice, just in front of White's cottage.

The strolling herd of 300 or 400 bunched just long enough to make this third hole—and then, incredibly, a fourth, not far from the one the men had approached. Mike got on the phone again, and diverted two of the trucks coming from the Home Ranch to go in by White's and empty that hole first.

To Mike's relief, Lawrence Graham and his crew soon arrived and the men began the arduous task of pulling the cattle out of the holes. Although the ice was a full 4 inches thick, it was a treacherous business. The ropeman walked as near to the edge of the hole as he dared and threw the soft rope over the head of a cow. At a nod from the ropeman, the seven- or eight-man team heaved the cow out onto the ice around the edge of the hole, being certain to slacken the rope before she choked. Most cows were too weak and cold to move at first and made no fuss when the rope came off; others charged whoever was near. The men would then move around to another side of the hole for the next cow, fearing that the weight of two together would cause the ice to break again.

Many of the cows that the men pulled out simply flopped on their sides. The death rattle of waterlogged lungs soon hailed the men from all around. The teams would break from pulling for a while to move those rescued yet still prostrate cows farther from the hole. Gradually as they thawed out, icicles dripping from their jaws, the cattle would lurch to their feet and walk off towards shore. In no time some began nosing in the snow around the lake's edge, gently tearing up mouthfuls of bunchgrass, as if nothing had happened.

The first hole was empty, 13 head saved, and the steadily enlarging crew moved to the next hole. They were pulling cattle out quite fast now. Meanwhile, the wandering herd had turned north again from White's and was heading directly across the ice towards the freshly emptied hole. Mike decided to see if he could turn the herd from the lake altogether.

He walked down towards the herd, past the gaping hole, and stopped. For a while the cattle kept coming, then the lead cow stopped and the others did too. Fearful that the smallest movement might make the cattle bunch up, Mike stood stock-still. He held his breath as he watched. Then one cow, 30 or 40 head back from the lead, spotted the 13 rescued cows that had already wandered up the sidehill nearest the road. She started towards them, straight across the frozen lake, taking the rest of the herd with her. As the last cow picked her way off the ice, Mike breathed deeply.

Ten head came out of the second hole, and the large crew moved over to the biggest hole of all, the one in the middle of the lake. These cattle had been in a long time and were even more weak than the first animals they had pulled out. Every so often, some men would take a few minutes to push the rescued cattle onto their chests, putting their legs under them.

Meanwhile, Neil, Arno and the Quarter Horse crew, and Stan Murphy's full truckload, had arrived at White's. There were now two holes there, containing about 20 head. The men began pulling. It took them some time to empty the two holes and then they walked the half mile up the lake to help at the big hole.

Halfway there, a big old high-horned cow came storming towards them, forcing the walking men to make a wide detour. She had already given a little trouble. Once out of the water she was so cold she could only stand, and though the men could see she wanted to charge someone, they knew she was too stiff to move and so went back to pulling. She began to ease her cold limbs, flexing first one leg and then the other. Then she chose her target. The pullers had just removed another cow and Neil Graham was on the end of the long rope. The horned cow lowered her head and knocked him down with a quick toss. Now she headed down the lake and on to Cayuse Mountain. No more ice for her!

With all the men working together, the cows were coming out faster. Just as the first men who had arrived on the scene were beginning to tire, Les Harris and his fencing crew arrived. Because their truck was without a radiophone, they had no inkling of the catastrophe. They had downed fencing pliers to lunch at Chapperon, but when they heard about the emergency from the cook, they had set out immediately to assist.

With the last cow out, the exhausted men stood up. Only a handful of animals still lay on the ice; the rest had wandered off the lake. It had taken roughly three quarters of an hour to pull 97 head out of the icy waters, every single animal that had been swimming when the men first arrived. And not one of those hardy Herefords suffered noticeably from her ordeal. (The crew of the 1 December disaster were amazed to see 9 head of cattle pop up to the surface of the holes later that month, and 15 more when the ice went off the lake the next spring. All were animals that had sunk below the ice when the initial groups broke through. For months after, anyone driving by would see a group of ravens fly up from the carcasses, or a coyote lope off to the hills.)

Thirty-seven pairs of hands had responded to Mike's call: men on the payroll, their wives who were not, and various friends and neighbours. The tired group began to go their separate ways, some warming up with coffee at White's, the wives returning to make lunch, others going to eat a late lunch and get back to the day's work. Lawrence Graham stayed until the last few head got to their feet and walked away.

The thought on Neil's mind as he drove back to the Home Ranch was that because of the rapid radiophone communication and the close-knit community that it made of Douglas Lake, 97 head were still alive. On the market that year, a cow was worth $300, which multiplied by 97 is $29,100, and the new radiophone system had cost $15,000. It had just paid for itself amply.

CHAPTER TWENTY-ONE

Douglas Lake is one of the unique cattle ranches of the world. It is almost completely self-contained—one of the few that is all in one piece, having bought properties to make it a more efficient operation. It has two commodities to look after: an abundance of beautiful grasslands that have never been abused, and an abundance of water. As long as beef remains an important food and the Government doesn't completely tax us out of existence, the ranch will go on for many years to come.

Many people have dedicated their lives to this ranch because they like the life. It is a way of life that is fast disappearing and anything I can do to maintain it, I certainly will.

Charles N.W. Woodward to the author,
13 February 1979

Both Chunky Woodward and Neil Woolliams understood well that in such years of rapid change, Douglas Lake owed its continued success in large part to the interest and devotion of its employees. As the steady crew shrank, each individual rose in importance. The job changed to fit the man rather than vice versa. Good wages and added responsibility rewarded good men.

Like Ward, Neil and Woodward appreciated that, generally, married men are reliable long-term employees, so whereas in the beginning there had not been one married man (or anyhow not one married woman) on the ranch, by 1976 there were 25, constituting over 50 per cent of the payroll. It was a side effect of the more obvious changes that had come about during Neil's management.

Few of the wives were of the type to be idle, and they too became involved in the running of the ranch. They cooked in the cookhouses; they stocked shelves and cut meat in the store, that centre of activity; they became cowboys during the peak periods of branding, turnout, fall roundup and weaning; they rode in the Howse Field looking for cattle down with ticks during the springs that Agriculture Canada conducted experiments to choose effective antidotes; they milked the small dairy herd and fed the orphaned calves; they worked in the vegetable gardens at the Home Ranch, Chapperon and Minnie Lake; they catered to the Panorama Sales; they raised chickens, ducks and geese. This pulling together by all the residents on the ranch helped in its unification and in the understanding of common problems.

Like most ranchers' wives, these were resourceful, talented women with hobbies to fill their spare time: leather tooling, beadwork, carding and spinning, macramé, embroidery, writing, painting, sculpting, professional cake decorating, guitar playing, and fly tying. Their pastimes took them outdoors: fishing, bird-watching, bird hunting, mushroom gathering, herb and Indian tea collecting, raspberry and huckleberry picking, cross-country skiing and horseback riding.

Three more of the original nine cookhouses closed down; the foremen's wives fed the single men in these locations. Arno Nielsen founded a Douglas Lake 4-H Club, helping teenagers rear, fit, and show calves. The women formed their own club, which began a library and later was instrumental in having the one-room school at the Home Ranch reopened. The Merritt School District had built it in the '50s to teach the children of logging families in the area, following Spencer and Ross's vast timber sales. The women organized Christmas, Easter and Hallowe'en parties for the children and taught each other their diverse skills at their monthly meetings. Television reception improved with better-located antennae and cablevision. As a co-operative effort, employees and management built a pool hall and a tennis court.

The Home Ranch crew put in a new water system to supply the growing row of homes, the cookhouse and several fire hydrants. The high water table around the Home Ranch provided a steady supply of rust-coloured water that turned the red-roofed, white-walled houses rusty from the lawn sprinkling. It also turned a whiskey and water black, which was more disturbing, especially to a loyal Scot like truck driver Scotty Paterson.

Modernization and camaraderie could not always provide the total answer to isolation. One winter, Grace LaFleure, the Minnie Lake foreman's wife, drove to Kamloops for a dental appointment. But she did not get home that night, nor for the next six, for strong winds and loose snow blocked the road, and government snowplows took a week of steady work to get it open again. Another winter, when Ken Allen was in charge at Minnie Lake, his Thompson wife, Liz, took young Johnny down to Quilchena over the drifts by snowmobile because he had a suspected ruptured appendix. A Merritt ambulance took him the rest of the way to hospital.

The presence of so many women on the ranch sounded the death knell for male supremacy in the cow camps. One cowboy left the crew for the precise reason that there were women other than the cook in camp.

The cowboy's role was now less arduous though more exacting than it had been in Joe Coutlee's day. Pickup trucks pulling horse trailers took the cowboys part way to their destination and all the way home on days of long rides when moving stock or changing camp. The job demanded less of the horses and was done with fewer of them; by 1978, the cowboy remuda was down to 222 mounts.

In other ways cowboying remained as ever. In all kinds of weather, in timber and open country, a cowboy calves out, ropes, wrestles, brands, dehorns humanely, turns out, scatters, rounds up, drives, cuts out, weans, and evaluates the cattle in his charge. He makes them move to a new field, to a corral, or into a truck quietly and gently, only occasionally resorting to the rattle of a can filled with rocks, the shock of an electric prod, the crack of a whip, or, more often, the urging of a well-timed "Hiya" and a little movement of his horse. He can spot a sick animal, diagnose its ailment, and treat it from the battery of drugs available. He keeps track of dead cattle and figures out the cause of death. He knows how to break, shoe, ride and care for his horses. He controls any cow dog he owns. He can wield an axe and a chain saw to build and fix fence and corrals, and thereby maintain Douglas Lake's high standard of fencing. He knows his range and how to salt it, and can identify and eradicate noxious weeds.

Cowboys and farmers are as easily singled out as ever. They are the men with a year-round tan from the V of their shirts to the brim of their Stetsons; they are the ones whose wives, even in the days of liberated women, deem it an honour to stumble around in the dead of night making breakfast. A Douglas Lake cowboy spends his free time learning to crack an Australian stock whip, team roping and calf roping at the Quarter Horse arena, practising his throw on a bale of hay, cleaning tack, making batwing (flared), stovepipe (fitted) or chink (knee-length) chaps, tooling and assembling saddles. A Douglas Lake farmer fills spare time gardening, tinkering with a vehicle, tossing horseshoes, playing tennis, shooting pool, playing his guitar, falling asleep in his chair, exhausted at eight each evening.

When the cowboys riding in Orval Roulston's Hatheume camp brought back the two-year-old heifers with their first calves in fall 1976, they were short 60 calves, 50 more than average losses. They did not pinpoint the cause until the next year, when it was clear that more than one killer bear was at work on the calf crop. Black bears—which include the black- and brown-coloured bears—are not by nature cattle killers, but once they get a taste of such a delicacy, there is no mending their ways. Moreover, one killer bear can teach her cubs the joys of killing livestock.

Hatheume riders found about a dozen badly mauled calves in their range early in 1977. They were obviously the victims of bear attacks. Their maggot-infested wounds caused them all to die despite the cowboys' doctoring. By the end of the season, 20 more calf carcasses had shown up, dead from the same cause. The regional predator control officer, Bob Lay, predicted that because of the ruggedness of the country, for every dead calf found another would lie undiscovered. Cattle counts after the fall drives proved him correct.

Various abortive efforts were made in 1977 to catch the killer bears. Officer Lay succeeded in poison-baiting one freshly killed calf, found at Mellin Lake, but the problem was too immense for him to solve. He advised Neil to cut down the bear population in the area between the Home Ranch and Hatheume the following spring until it became clear that he had killed the killers.

Orval Roulston, an extreme conservationist of game, understood the problem better than anyone. He also had time to do something about it, for Jerry McKenzie was now the cowboy foreman at Hatheume. Quelling his dislike of shooting, Roulston hired on as the ranch bear hunter, and in early May began prowling the timbered range he knew so well.

His successes came slowly at first, giving him ample time to salvage from each bear its salable spring hide for the fur market, its gall bladder for its medicinal, aphrodisiac and lure value, its claws for the reviving Indian craft industry, and its four paws for any Chinese delicatessen.

By mid-June, Roulston was averaging one bear a day and had shot 25. Of that number, one was a grizzly, the rest black bears, but he had seen three other grizzlies and the country was still alive with more blacks. There was no apparent explanation for the excessive bear population which Roulston slowly decreased.

Cougars can also be a problem with stock, as a Chapperon cowboy, Bill Brewer, knew. Two weeks after the Merritt conservation officer, Bud Ward, had shot a cougar that was killing heifer calves in the Hospital Pen at Chapperon, Bill Brewer rode home from doctoring calves that were wintering at Harry's Crossing. Hearing caterwauling and barking behind, he turned in his saddle to see another cougar attacking his blue heeler, Flicker. Bill sent in his other cow dog, Murph, to heel the cougar, and with the cat's attention distracted for an instant, he roped it by the head, dallied his rope on the saddle horn, dug his heels in his horse's flank and dragged the cat to the nearest fence, hanging the rope over the fence post.

The cat was starving and, it was later discovered, diseased. But Bill did not tell his wife, Rusty, what he was doing when he rode home for his gun until he had returned from the task, for she could not abide the killing of creatures, predators or others, and there was no time beforehand to explain how necessary it was to kill this cougar. Though a rope is an extension of a cowboy's arm, and Bill's reaction had been automatic in order to save his cow dog, his name was on everyone's tongue for weeks as the cowboy who had roped a cougar. Sadly, Bill Brewer drowned in Salmon Lake in August 1976.

Two-year-old heifers calved out at the English Bridge Maternity Barn for the last time in 1977. Orval Roulston, who had never been keen on running bulls with yearlings, had proved to Neil with carefully gathered facts that under their ranch conditions, such additional calf crops neither outweighed the loss of growth of their heifer mothers nor compensated for the shortening of the cows' lives. Three-year-old first calf heifers calved out at the Maternity Barn from then on.

Sam Bawlf, provincial minister of recreation and conservation, and Tom Waterland, provincial minister of forests and MLA for the area, accompanied Woolliams and Woodward around Douglas Lake's deeded land surrounding Courtenay, Crater and Pothole lakes one holiday weekend in 1977. The tour helped them to understand the ranch's increasing problems caused by visitors to these areas.

As many as 40,000 recreationalists now camped on that backyard of Douglas Lake during each tourist season. The all-terrain vehicles that these recreationalists brought with them—motorbikes and range rovers—dug up tussock after tussock of bunchgrass, irrevocably disturbing their delicate root structures. The drybelt range would take decades to recover from such damage. Unknowing campers chopped down the only ponderosa pine near a lake for the night's campfire; no longer would cattle shelter from sun or winds in its shade. Uncaring picnickers chased cattle away from lakeshores, stopping them from watering. With Woodward and Woolliams's blessings, Bawlf instructed his Kamloops staff to investigate turning the area into a provincial park, and trading this land belonging to Douglas Lake for Crown land more suitable for ranching.

In an effort to educate the agriculturally ignorant public, Neil urged Guy Rose, Joe Lauder, Gerard Guichon, the Upper Nicola Indian Band, and Jake Coutlee of Spahomin Cattle Company to join with Douglas Lake in preparing informative signs. Soon posted along the side roads, these signs told passers-by how delicate are the Interior ranges and how every traveller could help to preserve them for future generations by keeping all vehicles on the roads.

When a logging company headquartered in the Okanagan made overtures to build a logging road right through to Chapperon, Neil began to fight for the ranch's life. He succeeded only in obtaining a commitment that they would rip up the roadbeds once they finished logging.

The power line right-of-ways that had begun to crisscross the ranch earlier in the '70s had further opened up the timbered back country between Douglas Lake and the Okanagan Valley. These lines were coming from British Columbia's largest electrical substation, which B.C. Hydro and Power Authority was in the process of building on part of the Morton, Douglas Lake's best spring range. For many years the Hatheume cowboys had quit riding on Saturdays for fear of getting accidentally shot at by the many hunters who were enjoying their newfound access along these right-of-ways.

And this power that flowed across most of the Nicola Valley ranches in 230- and 500-kilovolt lines was not at first for local use but for use at the coast. All Douglas Lake's cow camps plus Minnie Lake were still without provincial power, though, ironically, when the Dry Farm became electrified along with Minnie Lake in June 1978, cook and crew alike were upset by the modernization. Even as pop-up toasters, plug-in kettles and electric irons started arriving at the Dry Farm, Neil was negotiating for the Boy Scout movement to electrify the summer camp at Sawmill Lake in return for the ranch's permission to hold their 1979 British Columbia and Yukon Provincial Jamboree on Alf Goodwin's old home place. The camp needed the power to provide the 2,500 scouts and their 600 leaders with running water and other amenities, so the Scout association agreed to install the mile of permanent line.

There was no such electrification in the cow camps at Louis's Corrals, Hatheume, Mud Lake, Raspberry Meadow and Courtenay Lake, so that flatirons, wood stoves, meat houses, outhouses and water running in the creek remained the most advanced conveniences at those scattered locations.

At Courtenay Lake each June, while turning cattle out to the Aspen Grove range, cowboys and their wives had nothing more than tents to sleep in for the month.

After the dry summer of 1977, which slowed both the hay crop and the cattle's conversion of grass to beef, Douglas Lake sold 1,000 extra calves in the fall, bringing the herd down to 10,000 head and the year's beef sales up to 3.85 million pounds.

A one-time Home Ranch mechanic, Norm Kato, and his crew—Donnie Earnshaw and Home Ranch wives—worked under a commercial fish farm licence in the name of Douglas Lake to net, gut and ship to coast markets 4,500 half-pound Kamloops trout grown in the private waters of Sabin Lake. This 1977 pilot project showed that a commercial fish ranch could be a valuable addition to a commercial cattle ranch.

With a steady crew of around 50 people, Douglas Lake went into the winter of 1977-78 having bought only an additional 700 tons of hay from the Spahomin ranchers. This purchase compared favourably with the average additional 3,000 tons ten years before.

The Home Ranch foreman, Arno Nielsen, expected to increase his hay tonnage in 1978 by another 400 tons grown on the newly levelled, plowed and seeded Home Flat opposite the Home Ranch. This flat, which a century before had been the site of John Douglas's cabin, had become the 200-acre site of Douglas Lake's first laboursaving, water-driven pivot irrigation system. In fact, had John Douglas's cabin still been there, the last wheel on the pivot would have trundled through the front door each time it neared the willows skirting the meadow's edge.

Gerard Guichon had installed one of the first pivot irrigation systems in Canada the previous year, but had not taken full advantage of its laboursaving potential. Ospreys nested in a tall ponderosa pine within the pivot radius so Gerard and his son, Laurie, left the tree standing and had to go out to the field every two days to reverse the long irrigation arm.

Buying by the truckload on a steadily strengthening market, Douglas Lake purchased, through Weiller and Williams Ltd., 3,000 yearling heifers in the spring of 1978, bringing the mature herd size up to 13,000 while the summer herd size rose again with the calf crop to over 17,000. Such cattle numbers had not been grazing on Douglas Lake range since the carefree years of the company's youth.

Consequently, Neil planned on marketing the most beef ever from the ranch that fall: 4.5 million pounds of North America's favourite red meat. It would leave the ranch as 3,000 yearling heifers for grass-fed beef, 2,400 yearling steers and heifers for feedlot buyers, dry cows for the fast-food industry's growing hamburger trade, veal calves and bologna bulls. Beef demand had certainly diversified.

What Chunky Woodward, Neil Woolliams and Mike Ferguson still demanded of the Douglas Lake stock, however, was what Brian Chance had demanded in the '50s: weight for age. They continued to look to the bulls for this fast growth potential. The crossbreeding of exotic bulls from continental Europe with Hereford females which caused such excitement in the late '60s and early '70s had passed by the more traditional commercial ranches, despite the boasts of growthier calves and heavier weaning weights. Crossing Black Aberdeen Angus bulls with Hereford heifers was a common practice and not part of the revolution. Douglas Lake first dabbled in Simmental crossbreeding, off the ranch. A few summers later the ranch rented ten Blonde d'Aquitaine bulls but found their offspring did not finish well on grass alone. Then Neil purchased four stretchy Simmental bulls from an Alberta rancher, Jerry Kaumeyer, in order to cross them on Douglas Lake stock and thereby raise more rugged half- or quarter-blood bulls. There was even talk of a controlled experiment using Maine Anjou, a breed which had evolved years ago from the French Mancelle crossed with Canada's old friend, the Shorthorn.

Meanwhile, it had been 25 years since Curtice Martin's Beau Donald Hereford bulls first mixed with the Douglas Lake herd, and Woolliams agreed with Woodward that a total change of Hereford sire bloodlines might enhance beef production faster. After searching extensively throughout Alberta in 1977, Woolliams and Ferguson came across the Three Hills ranches belonging to Wilf Evans, Allan and

Victor Garson, and Alton Parker. Their championship Hereford stock shared the same Canadian bloodlines, making their three herds almost interchangeable. The prospect of new blood mixing with Douglas Lake's fine 5,500-head brood herd intrigued both manager and cowboss. To ensure that the ranch would buy and keep only prime bulls, they introduced more annual checks: semen testing and scrotal measuring of each of the 300-head bull herd. Thereafter they looked down the road to fleshier and more vigorous "hundred and elevens" whose carcasses would yield a higher percentage of lean beef.

Neil Woolliams also saw a different horizon down the road. Douglas Lake had given its young manager a diverse education and a challenging responsibility over the years, but it had never removed the deep desire that surges within every wage-earning beefman—the necessity to go it on his own. In 1978, he and his family answered this call by purchasing a cattle and sheep property lying within Australia's Great Dividing Range. By the time they were to emigrate, Neil would have given the ranch 20 years of his life—high time for new blood to take over.

As Neil screened the candidates applying for his job, Douglas Lake Ranch rolled on towards its operational centenary in June 1984. The challenges facing the new, fifth manager of Canada's greatest ranch were many: maintain high ranch morale and thereby keep the good staff; roll with the punches doled out by government; preserve the integrity of the ranch and its way of life; slow the pace of expenses; and increase income to keep up with rising costs. If the range reseeding, cross-fencing, hayland fertilizing and improvement policies were continued, grass production would intensify; therefore beef gains would increase. Increased income should naturally follow. Thus even after a hundred years, Douglas Lake Cattle Company's best potential could lie in yet greater yields from its superb renewable resource: its grasslands.

EPILOGUE

The seven men who launched Douglas Lake Cattle Company were pioneers: adventurers with a strong sense of market and timing. They wove their accomplishments into the fabric of British Columbia's history. The memories of some of them live on; some do not.

Joseph Blackbourne Greaves's home town of Pudsey is now part of the city of Leeds. Michigan Bar, the California gold rush town where he made his first pitch as an entrepreneur, is now Ed Ruman's cattle ranch. His land near Savona is now part of the Indian Gardens Ranch.

Benjamin Van Volkenburgh and his family left no trace.

Joseph Despard Pemberton is remembered in Pemberton Point and Despard Cove on Broughton Island; the town of Pemberton, Pemberton Meadows and Pemberton Portage, all on the Douglas-Lillooet route north; Pemberton Road and Despard Avenue in Victoria. The Victoria firm of Messrs. Pemberton and Son which he founded in 1887 continues to do business as Pemberton Realty, Pemberton Labow Haynes and Pemberton Securities Limited.

William Curtis Ward was one of the most important bankers in early British Columbia: the Canadian Imperial Bank of Commerce continues to flourish, and is still the banker for the provincial government. A new Bank of British Columbia has received a charter, though it has no connection with the old bank of the same name.

Charles William Ringler Thomson left no descendants, and the Victoria Gas Company is now a subsidiary of B.C. Hydro and Power Authority. Ring-necked pheasants continue to have a hard time in British Columbia. C.N.W. Woodward released a flock of 600 of these birds at Douglas Lake in 1962, but only a few pairs withstood the predators and the cold winters.

Peter O'Reilly's home at Point Ellice, Victoria, has been preserved for the perpetual enjoyment of the public. Okanagan Indians, now intermingled with Thompson and Shuswap

Indians, Mexicans, Canadians, Americans and Europeans, still live on the Nicola Valley reserves, some of which O'Reilly created in 1868. In the 1970's the Indian land question in British Columbia surfaced once more, this time raised by the Indian themselves.

Although Charles Miles Beak made an enormous contribution to ranching in Canada—the initial amassing of the nucleus of Douglas Lake Cattle Company—only one small creek recalls his name.

Before the turn of the century, there were many large, western Canadian spreads but few have survived. Douglas Lake Cattle Company on the other hand has grown steadily in size, in sales and in importance, and has for some years attracted recognition as Canada's largest. This has happened through a period when ranch operating expenses have gone up 150 times—from $8,000 in 1899 to $1.2 million in 1976—while cattle prices have risen only 20 times.

Though twentieth century progress threatens, if the gods permit, Douglas Lake Cattle Company will continue to be one of the world's finest ranches. Its bunchgrass stands will continue to attract some of the world's best cowmen and farmers; the land will continue to produce many staunch characters and many millions of pounds of beef year after year.

APPENDICES

COMPOSITION OF DOUGLAS LAKE
CATTLE COMPANY IN THE 1970's

Cow Camps

Courtenay Lake
Dry Farm
Hatheume
Louis's Corrals
Morton Cabin
Mud Lake
Portland Ranch
Raspberry
Sawmill Lake

Hay Ranches

Chapperon
  Harry's Crossing
  Head of the Lake
  Sawmill Lake
Home Ranch
  English Bridge
  Quarter Horse
  Operation
Minnie Lake
Norfolk

Grazing Permits on Crown Land

Aspen Grove Unit
  Buck Lake, Loon Lake and
  Pothole Lake areas
  Community Field
  Horse Lake area
  Kane Valley and Davis Lake
  One Mile area
Douglas Lake Unit
  Green Lake and Hatheume
  Lake area
  Pennask Lake, Paradise Lake
  and Elkhart Lake areas
  Salmon River area
  Tee Range
  Hamilton Commonage—Douglas
  Lake Cattle Company portion
Lumbum Unit
  Wilson Field
Moir Burn Field
Nash's Field
Peterhope Unit
  Jack's Lake
  Lauder Meadow and Moir
  Reservoir
  Whiterock Field

DOUGLAS LAKE CATTLE COMPANY OWNERSHIP, 1884 TO PRESENT

Douglas Lake Ranch	June 1884 to July 1886	Charles Miles Beak Joseph Blackbourn Greaves Charles William Ringler Thomson William Curtis Ward
The Douglas Lake Cattle Company, Limited Liability	July 1886 to January 1892	Charles Miles Beak Joseph Blackbourn Greaves Charles William Ringler Thomson William Curtis Ward
	January 1892 to August 1910	Joseph Blackbourn Greaves Charles William Ringler Thomson William Curtis Ward
Douglas Lake Cattle Company Limited Liability	August 1910 to June 1914	William Curtis Ward
	June 1914 to August 1950	William Curtis Ward and family
	August 1950 to April 1951	Victor Spencer William Studdert
Douglas Lake Cattle Company Ltd.	April 1951 to June 1959	Victor Spencer Frank MacKenzie Ross
Douglas Lake Cattle Company (1959) Ltd.	June 1959 to September 1969	Charles Namby Wynn Woodward John Joseph West
	September 1969 to present	Charles Namby Wynn Woodward

INDEX